T0227618

SYSTEM DESIGN USING THE INTERNET OF THINGS WITH DEEP LEARNING APPLICATIONS

SYSTEM DESIGN USING THE INTERNET OF THINGS WITH DEEP LEARNING APPLICATIONS

Edited by

Arpan Deyasi, PhD
Angsuman Sarkar, PhD
Soumen Santra, MTech

First edition published 2024

Apple Academic Press Inc.
1265 Goldenrod Circle, NE,
Palm Bay, FL 32905 USA

760 Laurentian Drive, Unit 19,
Burlington, ON L7N 0A4, CANADA

CRC Press
2385 NW Executive Center Drive,
Suite 320, Boca Raton FL 33431

4 Park Square, Milton Park,
Abingdon, Oxon, OX14 4RN UK

© 2024 by Apple Academic Press, Inc.

Apple Academic Press exclusively co-publishes with CRC Press, an imprint of Taylor & Francis Group, LLC

Library and Archives Canada Cataloguing in Publication

CIP data on file with Canada Library and Archives

Library of Congress Cataloging-in-Publication Data

CIP data on file with US Library of Congress

ISBN: 978-1-77491-258-4 (hbk)
ISBN: 978-1-77491-260-7 (pbk)
ISBN: 978-1-00337-665-1 (ebk)

About the Editors

Arpan Deyasi, PhD
Department of Electronic and Communication Engineering,
RCC Institute of Information Technology, Kolkata, West Bengal, India

Arpan Deyasi, PhD, is presently working as an Assistant Professor in the Department of Electronics and Communication Engineering at RCC Institute of Information Technology, Kolkata, India. He has 15 years of professional experience in academics and industry. His work spans the field of semiconductor nanostructure and semiconductor photonics. He has published more than 150 peer-reviewed research papers, including journal, book chapters, and conference papers. He has already edited/co-edited six books published by publishers of repute. He is associated with different International and national conferences in various aspects and was general chair at FRCCD 2015 and ICCSE 2016. He is a reviewer of a few journals of repute as well as several prestigious conferences in India and abroad. He is a senior member of IEEE, secretary of the IEEE Electron Device Society (Kolkata Chapter), and member of IE(I), Optical Society of India, IETE, ISTE, etc. He is working as single point of contact of several online certification courses and activities in his institute (NPTEL, IIRS, Coursera, EDx, e-outreach, Internshala). He is working as a faculty adviser of the student chapter of the Institution of Engineers (INDIA) in the ECE Department at RCCIIT.

Angsuman Sarkar, PhD
Department of Electronic and Communication Engineering,
Kalyani Government Engineering College, Kalyani, West Bengal, India

Angsuman Sarkar, PhD, is presently serving as a Professor of Electronics and Communication Engineering at Kalyani Government Engineering College, West Bengal. He had earlier served at Jalpaiguri Government Engineering College and Kalyani Government Engineering College as Lecturer, Assistant Professor, and Associate Professor of the ECE

Department for more than 20 years. He received an MTech degree in VLSI and Microelectronics from Jadavpur University. He completed his PhD at Jadavpur University in 2013. His current research interest include study of short channel effects of sub-100 nm MOSFETs and nanodevice modeling. He is a senior member of IEEE, a life member of the Indian Society for Technical Education (ISTE), an associate life member of the Institution of Engineers (IE) India, and currently serving as the Chairman of IEEE Electron Device Society, Kolkata Chapter. He has authored five books, many book chapters, 70 journal papers in international refereed journals, and 57 research papers in national and international conferences. He is a member of the board of editors of various journals. He is a reviewer of various international journals. He is currently supervising six PhD scholars and has already guided five students successfully as principal supervisor. He has delivered invited talks, tutorial speeches, and expert talks at various international conferences/technical programs. He has organized IEEE international conferences and several workshops/seminars.

Soumen Santra, MTech
Department of Master of Computer Application,
Techno International New Town, Kolkata, West Bengal, India

Soumen Santra, PhD, is presently working as an Assistant Professor in the Department of Master of Computer Application, Techno International New Town, Kolkata, India. He has 11 years of professional experience in academics and industry. His work is in the field of machine learning, the Internet of Things (IoT), and medical image analysis. He has published more than 50 peer-reviewed research papers, including in journals, book chapters, and conferences. He is associated with different international and national webinars in various aspects. He is a reviewer of a few journals. He has achieved the best mentor of NPTEL online certification at his institute. He is working as a mentor and consultant for several industry-based projects and is also a faculty adviser of the Entrepreneurship Cell and Coding Trainer for the Placement Cell of the CA Department of Techno International New Town.

Contents

Contributors

Shalini Adhya
Arya Parishad Vidyalaya for Girls, Kolkata, West Bengal, India

Arighna Basak
Department of Electronics and Communication Engineering, Brainware University, Kolkata, West Bengal, India

Jenifar Begam
Department of Civil Engineering, Techno International New Town, Kolkata, West Bengal, India

S. Benila
Department of Computer Science and Engineering, SRM Valliammai Engineering College, Chennai, Tamil Nadu, India

N. Usha Bhanu
Department of Electronics and Communication Engineering, SRM Valliammai Engineering College, Chennai, Tamil Nadu, India

Mainak Biswas
Department of Electrical Engineering, Techno International New Town, Kolkata, West Bengal, India

Moitri Chakraborty
Department of Electrical Engineering, Techno International New Town, Kolkata, West Bengal, India

Anustup Chatterjee
Department of Mechanical Engineering, Techno International Newtown, Kolkata, West Bengal, India

Sayan Chatterjee
Senior System Engineer, Cognizant, Technology Solutions, Kolkata, West Bengal, India

Shruti Shovan Das
Undergraduate Student, Asutosh College under University of Calcutta, Kolkata, West Bengal, India

Santanu Dasgupta
School of Hospitality and Tourism, Sister Nivedita University, Kolkata, West Bengal, India

Arpan Deyasi
Department of Electronics and Communication Engineering, RCC Institute of Information Technology, Kolkata, West Bengal, India

Meera Joshi
Department of Mathematics, Aurora's Degree and PG College, Hyderabad, Telangana, India

Talha Khan
Department of Mechanical Engineering, Wichita State University, Wichita, United States of America

A. Krishnaraju
Department of Mechanical and Automation Engineering, PSN College of Engineering and Technology, Tirunelveli, India

Srinidhi Kulkarni
Department of Computer Science & Engineering and Information Technology,
International Institute of Information Technology, Bhubaneswar, Odisha, India

Sanghita Kundu
Department of Education, Institute of Education for Women, Kolkata, West Bengal, India

Akash Maity
Department of Electrical Engineering, Techno International New Town, Kolkata, West Bengal, India

P. Marisami
Lecturer, Department of Mechanical Engineering, Government Polytechnic College, Salem,
Tamil Nadu, India

Caprio Mistry
Department of Electronics and Communication Engineering, Brainware University, Kolkata, West
Bengal, India

Partha Mukherjee
Infiflex Technologies Pvt. Ltd., Kolkata, West Bengal, India

Soumyajit Pal
Faculty of Information Technology, St. Xavier's University, Kolkata, West Bengal, India

B. Prakash
Second Engineer, Synergy Marine Pt. Ltd., Singapore

Soumen Santra
Department of Computer Application, Techno International New Town, Kolkata, West Bengal, India

Angsuman Sarkar
Department of Electronics and Communication Engineering, Kalyani Government Engineering
College, Nadia, West Bengal, India

Sweta Sarkar
Construction Project Manager, University of Western Ontario, Canada

K. Sreenivasan
Lecturer, Department of Production Engineering, Thiagarajar Polytechnic College, Salem,
Tamil Nadu, India

S. Srinivasan
Deputy Manager, Product Engineering, Bharath Heavy Electrical Ltd., Trichy, Tamil Nadu, India

Rishabh Kumar Tripathi
Department of Computer Science & Engineering and Information Technology, International Institute
of Information Technology, Bhubaneswar, Odisha, India

Abbreviations

ADC	analog-to-digital conversion
AEC	architecture, engineering, and construction
AES	advanced encryption standard
AHMS	automated healthcare management system
AI	artificial intelligence
ASCI	application-specific integration circuits
BDA	big data analytics
BIM	building information modeling
BIoT	building internet of things
BLE	Bluetooth low energy
CAD	computing-assisted design
CNN	convolutional neural network
CPT	current procedural terminology
DC	direct current
DES	data encryption standards
DLP	discrete logarithmic problems
DSP	digital signal processor
DWT	discrete wavelet transform
EB	exabytes
ECC	elliptic curve cryptography
ECDLP	elliptic curve discrete logarithm problem
ECG	echo-cardio grams
EHRs	electronic-health records
EMR	electronic medical records
EPC	electronic product code
FAO	Food and Agriculture Organization
FNA	fine needle aspiration
GNSS	global satellite navigation system
HEI	higher education institutions
HER	electronic health records
IaaS	infrastructure as a service
ICD	international classification of diseases

ICU	intensive care unit
IDC	International Data Corporation
IFP	integer factorization problem
IIoT	internet IoT
IoT	internet of things
IR	infrared
IT	information technologies
KB	knowledge base
KNN	K-nearest neighbors
LARS	low altitude remote sensing
LDR	light dependent resistors
LPWAN	low power wide area network
M2M	machine-to-machine
MERS	middle east respiratory syndrome
MIoT	manufacturing internet of things
ML	machines learning
MRI	magnetic resonance imaging
NIST	National Institute of Standard and Technology
NLP	natural language processing
OAEP	optimal asymmetric encryption padding
PaaS	platform as a service
PHI	protected health information
PIR	passive infrared
PKC	public-key cryptography
RFID	radio frequency identification
RoR	rate-of-rise
SaaS	software as a service
SARS	severe acute respiratory syndrome
SIG	special interest group
SIM	subscriber identity module
SKC	secret-key cryptography
SSO	single sign-on
SVM	support vector machine
TCIA	the cancer imaging archive
UAV	unmanned aerial vehicle
UN	United Nations
VLSI	very large scale integrated
VPN	virtual private network

VR	virtual reality
WBAN	wireless body area network
Wi-Fi	wireless fidelity
WoT	web of things
WSN	wireless sensor network

Preface

The recent pandemic has brought about significant change resulting in the utilization of scientific knowledge for the benefit of mankind. The analysis of data obtained from various reliable sources has played an important role in this conclusion. Moreover, industry 4.0 needs optimization of human resources and infrastructure to enhance productivity, and in this context, the Internet of Things (IoT) opens the gateway for qualitative outcomes in different arenas like healthcare, environment, automation, security, integrated service, cloud computing, etc.

The field of IoT becomes more promising when data analytics is mixed with it, and therefore, a complete automated manufacturing/service sector can be implemented with a broader area of coverage. The present generation of academicians, therefore, need to collaborate with industrial representatives who are working in this area. This will not only prove beneficial for the industrial revolution, but academics will also find more food for thought. This type of collaborative approach will help reduce the gap between industry and academia, and therefore, augment the progress of society.

Nowadays AI/ML/IoT has become a boon to society. World leaders are funding huge amounts of money to these research fields. Even the Indian Government is encouraging and trying to fund these areas. Currently, in India, many startups are coming up with unique ideas of using AI in agriculture. As we know, agriculture is the second-most booming market worldwide. However, India is suffering from a lack of AI experts currently and many AI training initiatives by various institutions are ongoing. Most of these though are very theoretical and provide only textbook knowledge rather than a practical approach.

Practical approaches help learners understand the gaps and limitations in the current AI frameworks as it is still only a research field. Even though many initiatives are being taken to utilize AI in daily activities, they are mostly formulated through pattern search/match problems. The goal of AI goes beyond this. Hence, practitioners need to understand these gaps and limitations and focus on these areas rather than on how to write codes using available programming tools.

Knowledge in these areas should be focussed on understanding open areas and challenges in the current available frameworks and architecture, and exploration of alternative possibilities instead of training on a coding platform, solving linear equations on predefined datasets that have been solved 50 years ago.

The present edited book is an honest step towards the creation of a collaboration where both academicians and industrial workers make novel initiatives for solving the present-day problems in various industries and enhancing people's comfort in their homes. Several key and emerging areas like automobile, hydraulics, manufacturing, hospitality, home automation, industrial optimization, and most importantly, the COVID-19 pandemic situation is covered within these two volumes.

The results of the book are novel, and can safely be applied in their respective domains. Researchers have shown the simulative findings of their proposals, and comparative study reveals the supremacy of their findings compared to the existing technologies. In this context, the editors expect that the solutions presented herewith can meet the ever-increasing demand of industries, at least in a few sectors. More optimized solutions can be generated by another group of researchers/workers in this field, perhaps, in the near future. The editors thus hope that the present work is a valuable step towards the betterment of civilization.

Introduction

The Internet of Things (IoT) was created to meet the ever-increasing demand for connectivity and easy integration with other devices as well as virtual association with big data analytic (BDA) platforms. However, many hurdles exist in reaching this goal, including the ever-growing need for customer satisfaction. This involves data collection, design issues, application connectivity, implementation hurdles, storage, and analysis in a cloud-based environment. These associated features along with the IoT security and privacy problems require researchers to investigate optimized solutions while keeping track of the financial requirements.

It is already clear that the evolution in the domain of AI can neither be achieved by only computer engineers or neuroscientists. Many collaborative initiatives between engineers and neuro-scientists by world-renowned institutions have taken place to achieve this, and these initiatives are driving success in the use of AI towards society's use of IoT devices, engineering, etc. As a result, the editors believe that this book will be useful to academics as well as developers who are seeking AI-based solutions associated with a novel embedded system.

Healthcare applications that use IoT are used by both individuals and healthcare businesses in this era of pandemic, with a rise in the number of diseases worldwide. This necessitates the creation of analytical reports – a necessary component for the success of such healthcare applications. In particular, a large amount of data is generated that must be evaluated in order to produce the desired results. To do this, a cloud-based analytical solution will be the best option, as it will allow for speedier report generation than traditional methods.

Today, big data, and the Internet of Things (IoT) are two primary controlling realms. Every day, 2.5 quintillion bytes of data are created, according to estimates. Big data is a word that refers to a large volume of data in both structured and unstructured formats. Big data analytics is used in businesses and other application fields with high information density to produce predictions and forecasts. This helps businesses make better decisions in order to strengthen their operations.

Big data analytics is a technique for analyzing a large amount of data at once. Big data and the Internet of Things (IoT) were created on independent technologies; nevertheless, with time, they have been intertwined to improve their functionality. Since there is a scarcity of big data analysts, the data generated by the apps is not efficiently exploited and examined. For real-time IoT applications, higher synchronization between hardware, code, and interfacing is required.

The objective of this book is to produce comprehensive cutting-edge findings (either in prototype or model format) based on Internet-of-Things embedded with Deep Learning technology, which will be beneficial for people from all walks of life. The major focus is to create a space for an open AI/ML and IoT community across India and worldwide of doctors, engineers, researchers, and industry practitioners to develop products and create frameworks for addressing open challenges and gaps towards AGI for mankind. With this, we hope that this earnest endeavour will become a success and prove useful to readers.

CHAPTER 1

Applications of the Internet of Things and Big Data in Automated Healthcare

S. BENILA[1] and N. USHA BHANU[2]

[1]Department of Computer Science and Engineering,
SRM Valliammai Engineering College, Chennai, Tamil Nadu, India

[2]Department of Electronics and Communication Engineering,
SRM Valliammai Engineering College, Chennai, Tamil Nadu, India

ABSTRACT

A connected automatic healthcare system can play a major role in enabling patients to access and track their health data and permit for flawless communication with the providers of healthcare. The applications of IoT devices in the healthcare business facilitate doctors in observing patient activities and advanced treatment and give suggestions regarding patient's health with the decision tools and automation technologies. With the employment of this automatic technology, there are incomparable benefits that may improve the standard and potency of treatments and consequently improve the health of the patients with reduced cost. As the range of patients is increasing day by day, the healthcare business plays a crucial part in the generation of massive amounts of data. The amount of data is continually growing, and that comes from numerous sources; managing this information is extremely critical. Moreover, it is very challenging to retrieve this information and for further investigation. Big data analytics (BDA) is an alternative approach to healthcare systems that must estimate the affordable time for making sensible discretions organizing

System Design Using the Internet of Things with Deep Learning Applications.
Arpan Deyasi, Angsuman Sarkar, and Soumen Santra (Eds)

future views, and maximizing value. The analyzed information helps physicians to create choices for patients. In fact, healthcare BDA has the ability to lower treatment costs, predict epidemic outbreaks, avert preventable diseases, and improve people's overall quality of life. This chapter examines the need for IoT in the healthcare business, data analysis, and BDA. Investigations into various sorts of existing big IoT data analytics, as well as the relationship between IoT and data analytics in the healthcare industry, are also explained. The chapter also focuses on open challenges faced in the real-time implementation of connected healthcare and BDA.

1.1   INTRODUCTION

Mechanization has a major impact on the economy in the general global financial system, and in our everyday lives. Engineers are attempting to construct complex systems by combining automated devices with mathematical and organizational tools for a large number of applications. The use of digital control systems and their application in information technology minimizes the need for human labor in the production of goods and services, known as automation. The employment of computers in healthcare is referred to as automation, where various technologies increase productivity in the delivery of medical services. The use of automation in the form of control systems and advanced technology to eliminate or reduce the need for manual tasks is referred to as automated healthcare solutions. For the sake of performance, this technology makes use of modern techniques and equipment. The automated healthcare management system [1] (AHMS) is a computerized system that assists in the management and administration of patient data. They allow you to keep an eye on your loved one's safety, protection, and health as you can do it from the convenience of your own home. Many routine activities, such as data entry, record maintenance, and patient health monitoring, are part of the healthcare industry.

With the introduction of computers in recent years, clinical tests and medical records in healthcare systems have become more standardized [2]. Electronic medical records (EMR) are kept up to date in order to offer better care to patients and physicians. As a result, medical care providers would have even more accessibility to a patient's complete medical history and all of the patient's medical information. This involves medical detections, medications, and data related to them, such as aversions, enumeration, and diagnostic sample-test outputs. The component of the automated healthcare system is depicted in Figure 1.1.

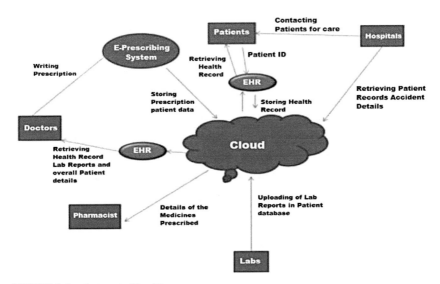

FIGURE 1.1   Automated healthcare.

There was a significant reduction in unnecessary and additional inspections as a result of this, and there is greater collaboration among various healthcare providers. Professionals in healthcare reported that they use internet-based and automated platforms to advance in their medical practice by using auto prompts for immunizations, cancer detection, and other periodic inspections.

Data is vital for any organization because it is often used to predict future behavior based on available data and current situations with certain parameters. In healthcare, administrative, and demographic data, diagnosis, care, prescription medications, laboratory testing, physiologic tracking data, hospitalization, patient insurance, and other data are all collected [3]. The most popular sources of health statistics are assessments, clinical and administrative reports, billing information, health statistics, inspections, diagnosis codes, and peer-reviewed literature. Automated healthcare includes patients in healthcare facilities, the number of physicians, nurses, and other medical professionals. These experts all keep track of patient care data and ensure that patients receive quality care and the best possible results. In today's dynamic environment, healthcare organizations manufacture immense knowledge that has several advantages and challenges and needs advanced technologies. Managing this immense knowledge could be a major task for organizations.

The study of current and past data related to the market in order to forecast tendencies can enhance services and even handle healthcare analytics which is the study of disease transmission. The industry sector covers a wide range of sectors and offers both macro and micro viewpoints. It has the potential to lead to improved patient care, clinical records, diagnostics, and business management. Another area where healthcare data analytics shines is in providing information to hospital administration that allows for the most efficient scheduling of physicians. In this situation, healthcare analytics provides a birds-eye perspective of physician reports and patient history, as well as the requirement to ensure that the most vulnerable patients are referred to the appropriate doctor or specialist. There are advances within the technology that helps in generating additional knowledge, which becomes unmanageable with the current potential. This creates a state of concern to develop new approaches so as to rearrange the information and extract useful patterns and apply this for future analysis.

1.2 TECHNOLOGIES USED IN AUTOMATED HEALTHCARE

Communication device computation, i.e., IoT, internet mobile Technology, cloud services, data analytics, 5G, nanoelectronics, and machine intelligence, as well as current biotechnology, are at the heart of smart healthcare [4]. In all facets of smart healthcare, these innovations are commonly used. The commonly used technologies in healthcare automation are shown in Figure 1.2. In Medical health-related issues, IIoT (internet IoT) plays a major significant role that benefits patients, caretakers, healthcare professionals, hospitals, and other connected organizations. In the case of patients, IoT has improved people's survival, especially for the elderly or those who live alone, by allowing for continuous tracing of routine activities or illnesses. In the healthcare sector, cloud computing improves productivity while lowering costs.

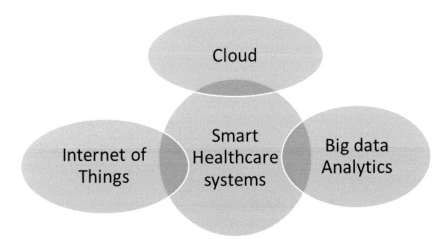

FIGURE 1.2   Automated healthcare with major technologies.

The exchange of medical records is made easier and more protected with cloud smart technology, which also mechanizes the methodologies with backend processes, and also allows the development and maintenance of telehealth apps simpler. It has given people the ability to collect, process, evaluate, and assimilate vast amounts of disparate, organized, and unstructured data generated by today's healthcare systems. Big data analytics (BDA) has increasingly been used to help in the delivery of treatment and disease research.

1.2.1   THE IOT AND ISSUES OF HEALTHCARE

The smart device-based IoT is a tool of interconnected, smart devices that can collect and transmit data across a wireless network without requiring human engagement. The opportunities for personal or professional development are limitless. Because of internet of things (IoT)-enabled equipment, remote access in the medical industry is now possible. The IoT has the ability to keep patients safe and secure while also helping doctors to provide exceptional care [5]. Interactions with doctors have also become simpler and more effective, which has improved patient involvement and satisfaction.

In IoT, medical devices can collect vital data and transmit it to doctors in real-time. Regardless of place or time, the reports offer an accurate assessment of the patient's condition. Patients would be able to communicate with experts with big thanks to technology, where doctors can now practice medicine from the solace of their own space-connected devices and wearables. IoT provides benefits to patients, doctors, and hospitals in automated healthcare systems.

1. **For Patients:** The IoT has improved people's survival, especially the survival of the elderly or those who live alone, by allowing for continuous tracing of everyday routines or illnesses. An individual or a concerned family member will receive warning signals if these activities change. Calorie counters, blood pressure, and heart rate can all be checked using exercise bands and other wireless devices.

2. **For Doctors and Healthcare Professionals:** Using IoT-enabled medical devices, doctors can keep better track of their patient's health. They will also see if any immediate medical attention is needed. Data obtained from IoT-enabled medical devices will assist doctors in determining the most effective treatment strategy for their patients and achieving the desired results.

3. **For Hospitals:** Hospitals can use IoT-enabled medical devices with sensors to check fridge and freezer temperatures, regulate drug inventory, and track the real-time location of ventilation tubes, mobility aids, nebulizers, and other hospital instruments.

1.2.1.1 THE BENEFITS OF IOT IN AUTOMATED HEALTHCARE

1. **Remote Monitoring:** Real-time remote monitoring via IoT Based devices and smart notifications can accurately identify illnesses, cure diseases, and save lives in the case of a medical emergency.

2. **Prevention of Diseases:** Intelligent sensors monitor health status, lifestyle factors, and the ecosystem in order to recommend precautionary measures that will lower the incidence of infectious diseases and traumatic situations.

3. **Reduction in Healthcare Cost:** The IoT eliminates the need for expensive doctor visits and hospital admissions, as well as making research more accessible.

4. **Medical Data Accessibility:** It allows patients to receive high-quality treatment while also assisting healthcare providers in making the best medical choices and avoiding complications.

5. **Better Treatment Governance:** IoT devices aid in the tracking of medication administration and treatment response, as well as the reduction of medical errors.

6. **Quality Health Management:** Healthcare organizations can acquire vital information on equipment and workforce productivity by utilizing IoT devices, which they can then utilize to make recommendations for improvements.

1.2.2   OVERVIEW OF BIG DATA

Big data is a word that refers to large volumes of data that can solve many real-life problems. It has piqued people's interest in recent decades due to the enormous potential it possesses. Various government and corporate industries collect, store, and evaluate data in order to improve their services. People working for different organizations all over the world generate vast amounts of data every day. The term "Big Data" [6–9] refers to the enormous quantities of data that are produced, replicated, and consumed. The International Data Corporation (IDC) predicted the digital data size to be at 130 exabytes (EB) in 2005. The digital universe reached around 16,000 EB in 2017. By 2020, according to IDC, the digital universe will have grown to 40,000 EB [3].

1.2.2.1   CHARACTERISTICS OF BIG DATA

The following characteristics can be used to characterize big data:

1. **Volume:** A significant amount of data is referred to as "volume." The amount of data plays a critical role in determining its worth.

When the amount of data is extremely high, it is referred to as "Big Data." This shows that the quantity of data in a collection decides whether or not it can be categorized as big data.

2. **Variety:** Structured, semi-structured, and unstructured data types are all content in that section by default. It may also apply to a variety of sources. The variety of data comes from sources both within and outside the company.

   Structured data is simply data that has been structured. It usually refers to data that has been specified in terms of size and format.

   Semi-structured data is data that is organized in a semi-structured manner. It's a kind of data that doesn't follow the traditional data structure. This form of data is represented by log files.

   Data that has not been organized is referred to as unstructured data. It usually refers to data that doesn't exactly fit into the traditional row and column structure of a relational database. Unstructured data includes things like text, images, and videos that can't be restored in the forms of tables or rows and columns.

3. **Velocity:** The term velocity refers to the rapid accumulation of data. Data comes in at a high rate from computers, networks, social networking sites, cell phones, and other outlets in high velocity. This determines how quickly data is produced and processed in order to meet demands.

4. **Variability:** It alludes to data inconsistencies and uncertainty, i.e., available data can become chaotic at times, and reliability and efficiency are difficult to monitor. Because of the numerous data dimensions arising from big Data can take many different shapes and come from a variety of different sources.

5. **Value:** Data is of little value or significance in and of itself; it must be turned into something useful in order to obtain information. As a result, one can claim the value and it is the most powerful of the five Vs.

1.2.2.2   REQUIREMENTS OF BIG DATA IN HEALTHCARE

In the healthcare system, big data is the data generated by the advent of digital technology that capture patient records and aid in the management of hospital outcomes, which would otherwise be too broad and complex for conventional technologies. Big data in healthcare refers to the gathering, analysis, and application of data that is too large or complex to be understood using conventional data processing methods. Machine learning algorithms and data scientists are often used to process big data.

1.2.2.3   HEALTHCARE BIG DATA AND ITS CHARACTERISTICS FEATURES

Automated healthcare is an interdisciplinary framework designed to prevent, diagnose, and treat human health issues or impairments. Medical professionals, healthcare facilities for supplying pharmaceuticals and other diagnostic or therapeutic technology, and a finance institution are all essential components of a healthcare system. Examples include hospital reports, patient medical records, medical test results, and IoT applications in healthcare big data outlets. Biomedical science generates a major amount of big data that is applicable to public health. Depending on the severity of the situation, different levels of treatment are required of the case. Professionals use it for primary care, as well as for acute care that necessitates the use of qualified professionals termed as secondary care, tertiary care such as advanced medical inquiry and treatment, diagnostic or therapeutic techniques. This data must be properly handled and processed in order to obtain meaningful information.

1.2.2.4   BIG DATA SOURCES FOR HEALTHCARE

Individuals are the source of all health data; however, the most direct source is informal personal data collection from a variety of sources. The sources of medical big data are given in Figure 1.3.

FIGURE 1.3   Big data in healthcare.

1. **Wearable Devices:** People can now monitor their heartbeat, cholesterol levels, weight, degree of activity, and amount of stress by using wearable devices.

2. **Apps on Smart Phones:** On smartphones, there are apps that monitor a user's workout routine and intensity, as well as the amount and quality of sleep they get.

3. **Sensors and Medical Devices:** Medical devices and sensors that can transfer data to the cloud include pulse dosimeters, insulin monitors, digital scales, heart rate monitors, Oximetry sensors, motion detectors, and future sensors that will deliver data from thousands of patients on a continuous basis.

4. **Electronic Medical Records (EMR):** EHRs are intended to give a more comprehensive view of a patient's medical information throughout lifetime.

5. **Electronic Health Records (EHR):** A patient's electronic health record (EHR) is a digitized representation of his or her medical records for paper document. EHRs are patient-centered, real-time databases that make information available to the public to authorized users promptly and securely.

6. **Insurance Companies:** Claims from private insurance companies and plans, claims from government health plans, and claims from pharmaceutical companies are all covered by insurance providers.

7. **Other Clinical Data:** Data received from medical imaging, research lab, pharmaceutics, health coverage, and other administrative data, as well as physician's written notes and prescriptions from computerized physician and clinical decision support systems.

8. **Search Engine Data:** On the internet, there are big data related to emergency care information, news feeds, and articles from medical journals.

1.2.3 BIG DATA ANALYTICS (BDA)

Applied to very large and complex data sets, BDA is the application of advanced statistical techniques. Data sets can include structured, semi-structural, and unstructured information, as well as information from a wide range of sources and sizes ranging from tens of terabytes to zettabytes in size. BDA is a type of advanced analytics that uses large amounts of data which entails complex applications that use analytics systems to power elements like statistical models, regression methods, and what-if analysis. BDA [8] systems and applications can help businesses make data-driven decisions that enhance business outcomes. Increased marketing efficiency, new sales opportunities, client personalization, and operational performance are all possible advantages. These advantages may provide strategic advantages over competitors with the right approach. Healthcare, consumer applications, manufacturing companies, banking and various technologies utilize BDA for future trend analysis and customer satisfaction. Figure 1.4 shows the usage of big data across different industries.

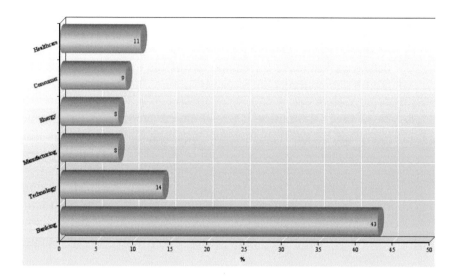

FIGURE 1.4   Big data analytics vs. applications.

Source: Peer Research - Big Data Analytics Survey

1.2.3.1   HEALTHCARE BIG DATA ANALYSIS

With clinical information, healthcare providers are able to dispense medications and make clinical decisions with higher precision, removing the guesswork that is sometimes associated with therapy and resulting in lower costs and improve quality of care. BDA has widely been used to help in the delivery of treatment and disease discovery. However, some fundamental problems inherent in the big data paradigm continue to obstruct adoption and research growth in this field. BDA is capable of enhancing patient-centered services, including early detection of disease outbreaks, novel insights into disease processes, and the monitoring of the effectiveness of medical and healthcare facilities. Big data is also in high demand in healthcare, owing to rising expenditures in countries such as the United States. According to McKinsey research, "with nearly 20 years of continuous growth, healthcare spending is predicted to account for 17.6% of GDP – more than $600 billion dollars more than the countries expected GDP and size."

The following are the number of services for healthcare data analytics:

- Determine daily revenues of patients so that staffing can be managed properly;
- Take advantage of electronic medical records (EHRs);
- Make real-time alerts available;
- Aid patients in cultivating a better relationship with their health;
- Make better educated strategic decisions by utilizing health data;
- Evidence-based medicine;
- Predictive analytics;
- Genomic analytics;
- Fraud reduction and improved data security;
- Use telemedicine to help people;
- Integrate medical imaging to obtain a more comprehensive view of the situation;
- Personnel management and smart staffing;
- Learning and growth;
- Advanced disease and risk management;
- Supply chain management;
- New cures and breakthroughs are being developed.

BDA is the analysis of big, diversified, complicated, dynamic, low-cost data sets in order to give tailored services. BDA's ability to convert data into data-rich judgments and deliver knowledgeable solutions to challenges in a variety of sectors. The use of BDA in healthcare has been investigated in several studies [6]. There has recently been a controversy in the literature over the use of BDA in healthcare [7]. Healthcare analytics solutions demand rigorous design and validation when patient safety, standards of healthcare, security, and evidence-based treatment are all on the line. While existing development techniques have been used to enforce solution design and analysis projects, it is important to consider how an effective extrapolation of the design correlates to more public understanding and observational method creativity. Researchers should be expected to rise in activities that enhance both research and clinical procedure.

1.2.3.2   CATEGORIES OF HEALTHCARE ANALYTICS

There are four categories of health analytics: descriptive analytics diagnostic analytics, predictive analytics, and prescriptive analytics. Figure 1.5 shows the different types of healthcare analysis techniques.

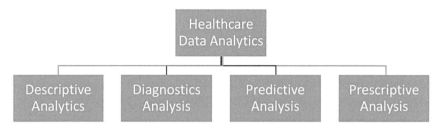

FIGURE 1.5   Types of data analytics.

While they're all interconnected and build on one another, therefore I grouped them into groups. The level of complexity and resources required rises as you proceed from the most basic to the most complicated types of analytics. Simultaneously, the level of added value and comprehension rises is shown in Figure 1.6.

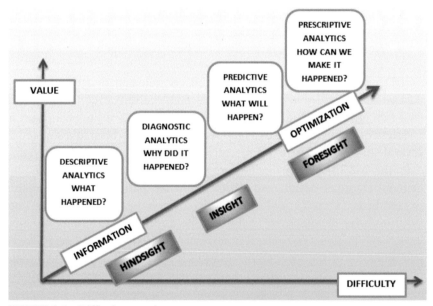

FIGURE 1.6   Difficulty vs. value.

Source: Adapted from Ref. [12].

1. **Descriptive Analytics:** It is the initial sort of data analysis. It is the basis of analysis of all data. In today's corporate world, it is the most basic and pervasive application of data. Descriptive analysis

provides a solution by offering a response to the inquiry "what happened" to the previous data in the form of dashboards. Descriptive analytics is used to investigate the effects of various healthcare decisions on quality of service, clinical trials, and outcomes [10].

2. **Diagnostic Analytics:** It is a subset of business intelligence that examines data or materials to determine the cause of an event. Some of the methods employed include drill-down, data exploration, data mining, and associations. Analysts pick data sources that will aid them in interpreting the outcomes during the discovery process. Drilling down entails concentrating on a certain aspect of the data or a certain widget. Data mining is a process for retrieving useful information from vast amounts of unstructured data that is performed automatically [11]. Finding consistent patterns in data can also help to accumulate information.

3. **Predictive Health Analytics:** Because they concentrate on the usage of data rather than just statistics, predictive health analytics are more complicated than simple descriptive analytics. It uses historical data and indicators to forecast future performance. When compared to traditional business techniques, predictive analytics using large datasets helps to improve the customer experience while also enhancing the outcomes. It aids in the analysis of massive volumes of transactional and unstructured data at once, resulting in conclusions that aid in the prediction of the future. In the event of an epidemic illness breakout, a pharmacist may need to forecast how much of a medicine to stock. Patients' length of stay; patients who might choose surgery; people for whom surgery is unlikely to be beneficial or who might suffer difficulties might all be anticipated and studied using huge amounts of previously collected data.

4. **Prescriptive Analysis:** It was named one of the hype cycle's technological innovations by the Gartner Group in 2013, and it has since gotten a lot of attention in the corporate sector [12]. Prescriptive analysis is a type of data analysis methodology that provides predictions and perspective information. The use of hybrid data, which includes both qualitative and quantitative data types, combining forecasts and prescriptions, taking into consideration

all possible complications, and employing evolutionary algorithms that can be easily customized to each case, in addition to the need for consistent and effective feedback, all are critical to the success of prescriptive analytics. Because they focus on the use of information rather than just statistics, predictive health analytics are more complicated than simple descriptive analytics. Descriptive analytics uses data visualization extensively to answer specific queries or find treatment patterns, giving a more complete view of evidence-based clinical practice. They allow businesses to manage operational content (info that is updated in real-time or near real-time) as well as collect all sensory information from people. This methodology can assist in establishing a better balance of capability and cost by detecting previously unnoticed trends in patients, such as tendencies associated with hospital hospitalizations [11].

1.2.3.3   OTHER TECHNIQUES FOR DATA ANALYSIS IN HEALTHCARE

1. **Fusion and Integration of Data:** The insights are more efficient and perhaps more accurate than if they were created from a single source of data by utilizing a collection of techniques that analyze and integrate data from numerous sources and solutions.

2. **Data Mining:** It is a typical approach used in BDA to extract patterns from massive data sets using a combination of statistics and machine learning methods inside database administration. When consumer data is mined to discover which segments are most likely to respond to an offer, here is an example.

3. **Machine Learning:** which is well-known in the field of artificial intelligence (AI), is also utilized for data analysis. It is a branch of computer science that uses computer algorithms to generate data-driven assumptions. It makes forecasts that human analysts would be unable to make.

4. **Natural Language Processing (NLP):** This data analysis tool, which is a subfield of computer science, AI, and linguistics, employs algorithms to analyze human language.

5. **Statistics:** Within surveys and experiments, this methodology is used to gather, organize, and interpret data.

1.2.3.4 CHALLENGES OF BIG DATA ANALYTICS (BDA) IN HEALTHCARE

Methods for managing and analyzing big data in healthcare are constantly being improved, particularly for real-time data streaming, acquisition, integration, and analytics utilizing Machine Learning and predictive methods. Electronic medical records (EMRs) can be better utilized with the help of data visualization in healthcare. Valid information gathering, security, data standardization, data storage, and data transfers are the significant obstacles that the healthcare business faces. Governance and ownership concerns, data analyst skill gaps, data errors, quality assurance, and real-time analytics are all examples of managerial challenges. The vital challenges faced by the healthcare industry in real-time implementation of BDA are discussed in further sections.

1.3   UNSTRUCTURED AND UNRELIABLE ORIGIN OF DATA

Patients generate a large amount of data that is difficult to acquire and manage using typical EHR formats. When it comes to handling massive data, the adoption of EMR becomes tough, especially when it comes to healthcare providers without a solid data organization. For clinically relevant information, including claims, billing, and clinical analytics, a single code must be maintained. To express the essential clinical ideas, medical coding systems such as current procedural terminology (CPT) and International Classification of Diseases (ICD) standards were developed [13]. Inconsistent formats, unprocessed and unorganized datasets, and a lack of transparency are some of the major obstacles in dealing with data. Cleaning of data on a regular basis can be done manually or automatically using a variety of customized algorithms to achieve high levels of correctness and integrity.

1.3.1   QUALITY OF DATA

Another problem in large data analytics is data diversity. Big data in healthcare is considerably less informative using traditional technologies due to its enormous size and very heterogeneous character. High-performance computing clusters connected to grid computing infrastructures are the most frequent platforms for running the development environment that aids big data processing.

1.3.2   STORAGE

One of the main issues is storing massive amounts of data, yet many businesses are comfortable with storing data within their own facilities. Controlling security, access, and availability are just a few of the benefits. The use of IT infrastructure for cloud-based storage is a better solution, which most healthcare organizations have chosen. Organizations must select cloud partners that are aware of the importance of healthcare-specific compliance and security concerns. Furthermore, cloud storage has reduced upfront costs, faster recovery procedures, and is easier to expand. Corporations can also take a hybrid framework for data storage, which may be the most adaptable option for providers with variable access to data and storage requirements. The technologies of wireless edge computing, cloudlets, and fog computing can be used in IoT to perform big data processing and analytics closer to the data source.

1.3.3   SECURITY OF DATA

To protect data from hackers and spoofing attacks, healthcare firms make data security as a top priority. Found traces of a number of vulnerabilities, a checklist of technical safeguards for protected health information (PHI) was created. Such standards, known as Health Information Privacy and Security Rules, assist enterprises with data storage, transfer, authentication mechanisms, and access, integrity, and auditing controls. Anti-virus software that is up to date, firewalls, encrypting sensitive data and multi-factor authentication are all common security precautions that can save a lot of frustration.

1.3.4   IMPACT OF ETHICAL AND LEGAL CONSIDERATIONS

The use of Big Data in healthcare poses significant ethical and legal considerations based on the subjective nature of the data contained. The risk of compromising individual privacy autonomy, as well as the influence on the public's demand for transparency, confidence, and tolerance, are all moral and professional problems when using Big Data. As a result, any institution that uses Big Data in the health sector must create affirmative regulations to protect personal health data in terms of confidentiality, security, and privacy, while simultaneously guaranteeing that scientific discoveries benefit from open data usage for the community's benefit.

1.4 CONCLUSION

This chapter presents an overview of IoT and Big Data applications in connected healthcare systems, with a particular emphasis on the use of data analytics. The importance of big Data analysis for maintaining the database and records of patients with varied diseases being discussed. The principle of big data in smart healthcare systems depends on various information, and this information can be used for predictive analysis of the occurrence of pandemic events. A large amount of medical data is dispersed over a variety of different platforms requires engineered solutions for integration of multiple system information and its implementation. The use of business intelligence in integrated healthcare system can be used for prediction of pandemic diseases, early warnings of occurrence of diseases thereby therapeutic solutions can be provided to raise the standard of human life.

KEYWORDS

- **automated healthcare management system**
- **big data**
- **big data analytics**
- **electronic health records**
- **exabytes**
- **healthcare**
- **Internet of Things**

REFERENCES

1. Suleiman, A. Y., Lydia, J. J., Kehinde, A. H., Kareem, A. A., & Abdullahi, Y., (2019). Development of an automated healthcare record management system. *Adeleke University Journal of Engineering and Technology, 2*(2), 79–90.

2. Sivasankari, A., Sindhuja, N., & Selvakani, S., (2019). Automated health care management system using big data technology. *International Research Journal of Engineering and Technology (IRJET), 06*(04).

3. Sabyasachi, D., Sushil, K. S., Mohit, S., & Sandeep, K., (2019). Big data in healthcare: Management, analysis and future prospects. *Journal of Big Data.* https://doi.org/10.1186/s40537-019-0217-0.

4. Shuo, T., Wenbo, Y., Jehane, M. L. G., Peng, W., Wei, H., & Zhewei, Y., (2019). Smart healthcare: Making medical care more intelligent. *Global Health Journal, 3*(3), 62–65.

5. Stephanie, B. B., Wei, X., & Ian, A., (2017). Internet of things for smart healthcare: Technologies, challenges, and opportunities. *IEEE Access, 5.*

6. Harerimana, G., Jang, B., Kim, J. W., & Park, H. K., (2018). Health big data analytics: A technology survey. *IEEE Access, 6*, 65661–65678. doi: 10.1109/2018.2878254.

7. Galetsi, P., & Katsaliaki, K., (2019). A review of the literature on big data analytics in healthcare. *Journal of the Operational Research Society,* 1–19.

8. Kitchin, R., (2014). The real-time city, big data and smart urbanism. *Geo Journal, 79*(1), 1–14.

9. Dinov, I. D., Heavner, B., Tang, M., Glusman, G., Chard, K., Darcy, M., et al., (2016). Predictive big data analytics: A study of Parkinson's disease using large, complex, heterogeneous, incongruent, multi-source and incomplete observations. *PLoS One, 11*, e0157077.

10. Raghupathi, W., & Raghupathi, V., (2014). Big data analytics in healthcare: Promise and potential. *Health Information Science and Systems, 2*(1), 3.

11. LaValle, S., Lesser, E., Shockley, R., Hopkins, M. S., & Kruschwitz, N., (2013). Big data, analytics and the path from insights to value. *MIT Sloan Manage Rev., 21.*

12. Linden, A., & Fenn, J., (2003). *Understanding Gartner's Hype Cycles.* Tech. rep., strategic analysis report, R-20-1971. Gartner, Inc., Stamford, CT, USA.

13. Roberta, P., Corrado De, V., Giuseppe, M., Katrin, G., Ilona, B., Walter, R., & Stefania, B., (2019). Benefits and challenges big data in healthcare: An overview of the European initiatives. *European Journal of Public Health, 29*(3).

CHAPTER 2

IoT in the Construction Industry

JENIFAR BEGAM[1] and SWETA SARKAR[2]

[1]*Department of Civil Engineering, Techno International New Town, Kolkata, West Bengal, India*

[2]*Construction Project Manager, University of Western Ontario, Canada*

ABSTRACT

Internet of things (IoT) is a significant emerging topic of economic significance, social significance, and technical significance. Our daily products, like durable goods, consumer products, cars and trucks, industrial components, sensors, etc., are being paired with robust Internet connectivity capabilities of data analytics for a smooth operation that will transform our daily life, and the way we work, lead, and move. The concept of combining computers and networks for daily devices like monitors and mobile phones has existed for decades now. Not only modern technology and market trends, but also several other industries are adopting IoT in their daily work process for the more smooth progress of work, thus bringing the IoT more deeply to the real world. One such large industry is the construction industry. More than about 16% of the nation's working population depends on its livelihood in the construction industry. Over 30 million people employ under the Indian construction industry, and it creates assets worth more than 300 billion. It almost contributes about 5% to the GDP of the country and 78% of gross capital creation. It is beginning to attract huge attention now a day in the construction industry. IoT devices, mobiles, and sensors are collecting job site data in a more efficient, affordable, and effective way than ever imaginable, with ease in

System Design Using the Internet of Things with Deep Learning Applications.
Arpan Deyasi, Angsuman Sarkar, and Soumen Santra (Eds)

work. In this recent concept, IoT is advancing the possibilities in various areas of application for smart construction. IT-based construction works are one of the prime and one of the most conservative industries in terms of adapting new methods or technologies. From a civil engineering point of view, we can assure that the use of IoT in the construction industry will simplify, to a considerable degree, the work. This chapter enlightens the knowledge of using digital technologies to transform construction work into a smart one. The fields where IoT can be applied for smooth as well as the rapid advancement of the construction procedure are discussed herein. There are actually huge challenges identified via some brief literature review, and the solution to them is delivered. In this era of digitalization, it's high time we apply IoT in our construction industry. At the moment, emerging countries, such as India, need rapid infrastructure and building growth. The solution to it aims to employ IoT technologies to accelerate building. Not only that it will make the industry smooth, but it will also help in increasing employment, business, and enrollment, and as a whole, the economy of our country. We have also gone through the feasibility of IoT so that the construction industry remains the largest industry in this world.

2.1   INTRODUCTION

The IoT links intelligent devices in your house and clever devices outside your home. The intermediary is mostly cloud-based, the link. Some building tools and appliances are clever enough to be monitored with a cloud app, and more businesses are looking for a way to build IoT and the cloud [1].

"The IoT enables objects to be sensed or remotely controlled via the current network infrastructures, providing ways to integrate the real environment in computing networks, improving performance, precision, and cost-effectiveness, in addition to reducing human interference," said the Global.

As IoT is expanded with sensors and drive units, the platform becomes an example of cyber-physical networks that also include technologies such as intelligent grid systems, automated power plants, intelligent houses, intelligent mobility, and intelligent cities.

Unmanneed aerial vehicles (UAVs) are benefiting from their popularity as modern vehicles. Monitoring and monitoring of huge building projects covering enormous areas, particularly drones, are facilitated by UAVs [2–4]. In addition, automated dump trucks and excavators are tested for the purpose of limiting human vulnerability to dangerous working conditions in different projects. Examples include Autonomous TMA Truck, Smart Trucks by Komatsu. Manual monitoring of vital equipment status and position on a building site requires a lot of time and is susceptible to human error. The design/project manager is very comfortable with fitting trackers for these vital properties. The system monitoring system IoT facilitates the user management, cost control, and decision-making of smarter equipment for building firms. The collection of documentation data and devices does not provide operational details easily to administrators [3–5]. IoT's beauty is that smart wireless networks are a cost-effective choice, even for small businesses and short-term ventures.

The construction industry remains among the last sector to apply the internet of things (IoT) in its processes and activities. But as already known, "better late than never," the industry has started implementing IoT in various aspects, such as site monitoring, machine control, fleet management, and so on. The latest estimates suggest that the time before IoT becomes a must-have tool in the building is only a matter of time. Emerging innovations can help to improve waste disposal [6] and the well-being of employees, reduce building costs, and help properties fit prospective customers. IoT uses internet sensors that are generally installed around work sites or can be used by workers. These IoT construction devices can collect data on construction activities, their result, and conditions and send it to the central dashboard in which to inform decisions the data is analyzed [7].

The sector is seeing an increase in operating complexity, and construction firms have grown and expanded globally and in the form of projects they perform since the beginning of this decade. It requires project managers in different locations at the same time, thus optimizing asset life cycle management and managing programs in order to minimize additional costs and to avoid future delays. Due to the use of digital solutions such as worker wear sensors and hardhats, wearables that track dangerous building materials, sensors that monitor and alert workers should enter insecure locations, IoT is able to make places more secure by connection on the construction site, for instance in construction with the ability to substantially minimize any kind of accident [8].

Effective reporting is possible by the connection through monitors, CCTV cameras, and even drones to construction sites all data are redirected to a central office, where recommendations about where a building project is located, and the changes required to keep the project underway are taken. Material sensors and RFID tags can facilitate automatic workflows, constructive material purchase and equipment maintenance. Sensors on-site facilities often track levels of use and may highlight possible problems in proactive maintenance – buildings that lack supplies or need maintenance for equipment previously entailed expensive delays. IoT assists in maintaining programs [9, 10].

The use of IoT devices could increase the control of resources at building sites significantly. If one has chips connecting all your computers, personnel, and supplies to the internet, one can automatically find them. There are substantial future costs and gains in time. IoT instruments will constantly provide updates on the conditions both at constructed and at under-construction locations through sensors strewn over a construction site. Sensors may detect such things as irregular movements on a machine piece which means that it needs to be repaired. One can identify moisture increases, which can alert the inspection teams about moisture problems. The construction industry being highly saturated, construction project managers and engineers need to beat the market competitors by both quality and time. In the case of both built and sub-construction places [11, 12], IoT instruments are continuously updated through the sensors on the building site. Sensors can sense unusual motions on a part of the system, meaning it must be fixed. High water levels can be identified and can alert the inspection teams of issues with moisture.

Computing-assisted design (CAD) helps people to transcend physics and drawings to a digital world where 2D or 3D models, structures, and more are created. The CAD and IoT merger of virtual reality (VR) headsets are a growing example. They complement the workflows of engineers by allowing quick returns to save time and resources [13].

Related assistance is offered in the architecture industry through building information modeling (BIM). Until instructing teams to assemble them, the project managers will evaluate simulated scaffolding configurations or let customers see computerized constructions before they do. BIM makes it possible for all stakeholders to enhance coordination and teamwork. Entities support the strategy of programs before they break through, frequently proposing new ways of doing their work and seeing if their positions lead to ultimate progress. In engineering and similar fields, CAD and IoT were not always commonly used together [11–13]. This is evolving today, and for the BIM and IoT there is a similar change. One choice is to import IoT sensor information to supplement BIM content. If the sensor displays current energy demand, building management will use those numbers to assess how they can change effectively and exclude those who would not perform properly. Builders can perform better than ever, reduce the number of people required for complicated tasks, and lower labor costs, with the use of IoT-enabled teamwork tools. For example, drones that are installed with advanced 3D mapping capability will search the job site and feed the information into a BIM device that produces exact 3D work models. These templates can be easily revised by the job manager and notified of the updates immediately to both employees and managers. It also helps stakeholders to collect up-to-date progress updates and videos of the building site by capturing data on the work site and posting them in real-time to an IoT website, enables them to monitor the status of a project and to make changes, if necessary, to monitor the project [14].

Even if 87% of shoppers take a green call, market research reveals that only 33% allow green shopping successfully. For two main reasons: cost and convenience, the purchasing behavior does not satisfy the consumer's request. When it comes to a big-ticket construction initiative, eco-building events are the perfect distribution channels. And IoT allows the public to deliver what they need. The environmental movement caused a flurry of interest about greener building activities. The IoT advantages are built to create a smart facility which knows when the construction is out of business. Automatically shut down unnecessary processes that save electricity and natural resources. Smart window designs provide a smarter operation than basic remote control or blinds that are time activated [15]. As IoT touches windows, it results in a device that detects natural sunlight or temperature parameters, and automatically closes lathes to save energy as things get cold or start to heat up.

The results of a building project depend primarily on whether the materials necessary for the completion are available [16]. A lack of funds could hinder progress, but machinery that collapses or gets robbed could also become misplaced. It avoids such catastrophes by using the IoT in building. Managers can also get data on different platforms in real-time. If a team leader uses an IoT system to monitor a geofencing element, virtual borders for tools and equipment can be created. Such features prohibit anyone from leaving something inadvertently following a changeover or from deliberately misusing equipment by removing it from the site.

Confidence in IoT will also help in the control of resources by alerting the public of anomalies until they trigger faults in the machine. Temperature, fluid speed, acceleration, and more information can be collected by sensors, enabling managers to respond more effectively to signs of problems. In connection, if an operator of an equipment finds something odd and reports it to an operator, he can promptly examine the problem by carefully examining the related information.

Artificial intelligence (AI) with its subset machine learning (ML), these two have their capacity to manipulate data, can also play an alarming role in construction and make it ideally suitable to the industry, becoming a tangible benefit. Building firms may use of AI and machine learning for monitoring information on stock level, scheduling information, and even on data obtained for projects, information on weather or other disturbances, to help advise and prevent possible errors or problems in previous projects. They allow you to properly evaluate time frames and get a better view of the amount of people and other resources available for projects [17].

Technology may make advances on properties and programs based on analytical insights from AI or machine learning so that it learns that it can take precautionary steps as parameters shift 'automatically' (Figure 2.1).

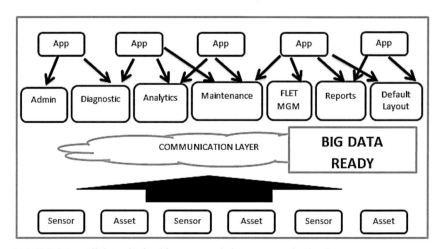

FIGURE 2.1   All data obtained is processed via a communication layer.

The IoT is staying here. It's the future, and it's the present for many. In the building industry, consumer demands have been boosted by the success of technologies that allows builder contractors to more effectively offer projects based on their own historical data and to achieve their tasks quicker, more reliably, and efficiently using intelligent embedded technology [15, 17].

Data had previously been obtained manually from sensors and thresholds crossed were frequently not detected until the harm was done first. Products that track ambient conditions at a low cost and warn when limits are crossed have been available in real-time. In order to ensure the surface conditions, satisfy the floor manufacturer's specifications, there are also options for measuring the humidity and temperature of cement floor plates as well.

Monitoring of the worksite is another increasing IoT application. Site administrators will track their activities from a single dashboard by equipping employees with low-powered IoT sensors. The sensors capture and relay data visually, which alerts monitors to unhealthy patterns of activity or efficiency delays. The potential to capture large volumes of sensor data is one of the leading IoT applications for building and the program turns the knowledge obtained into operating observations that nothing can compare with speed and precision.

Constructing new infrastructure in crowded urban areas also involves working together with existing structures that may be affected by development. There are now automatic deformation control systems, which can provide data 24/7 until configured. Intervals of measurements historically listed in days can now be set to minutes between periods. If the deformation levels determined by the record engineer are violated, staff is automatically notified and work may be halted and remedial steps taken, depending on the situation. Coordinated work on crowded roads will provide practical problems for the events that affect public right of way in congested urban environments. This refers to the completion of such activities beyond regular operating hours. However, as more people choose to not only work but live in the center of town, it is difficult to do this job without interrupting residents. Solutions will now not only warn us to exceed the threshold, but also of the importance of monitoring noise in order to determine the root problem and take appropriate action.

IoT may help decrease waste generation in building. Any industries also use sensor systems to track the garbage level on site. Certain devices can minimize cutting or molding waste and debris. There is also a program for waste control to measure the exact number of materials used to avoid waste generation. Finally, third-party agencies provide resources for the collection of data in real-time to help companies deal more effectively with their waste [8–10].

IoT data may support companies analyzing and improving their operational processes to further maximize performance, productivity, and, as a result, profitability on a larger, strategic scale. Given how IoT is transforming companies in all other industries, there is no doubt that IoT is transforming companies from construction and the industry. Building firms must, however, find ways to make use of IoT and participate in pilot projects to benefit from the early adoption.

Poor maintenance activities also lead to higher project completion costs. Outdated technological technologies don't make it possible for company owners to effectively track all assets, control equipment efficiency and prepare up-to-date upgrades, plan maintenance stops. The integration of IoT into building provides administrators with a continuous stream of real-time data that is handy for all types of decision-making — job planning, distribution of resources, monitoring, investment negotiations, etc. Access to the related data on request will enhance coordination between teams, protect teams against unfounded decisions, and enable asset monitoring in real-time [8–11].

Unforeseen modifications are characteristic of a project of building. In addition to helping project supervisors see unexpected problems in the corner, the IoT gives them an insight into how to respond. Although IoT is not vaccinating an organization against a flat-footed capture, the potential for a surprise to erase a carefully planned strategy can be significantly reduced.

2.2 DEFINITION AND ARCHITECTURE OF INTERNET OF THINGS (IOT)

According to a group of research projects in Europe, they quoted, "IoT is a unique part of perspective internet and may be regarded as a dynamic global collection of connections with self-armature potentials customary for standard and interoperable messaging etiquette where physical and implied "things" have an identity, material impute, with implicit disposition as well as the smart interface, which are completely integrated into the data network" [4–6, 8] (Figure 2.2).

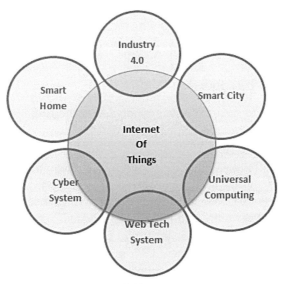

FIGURE 2.2   Consents of the internet of things.

2.3   TECHNOLOGY, SOFTWARE, AND HARDWARE

Before defining the technology, it is critical to analyze the vision of the IoT, which must fulfill the interaction between humans and the environment. The IoT architecture should be built on top of a network frame that integrates cabled and mobile technologies in a transparent and unified manner. Radiofrequency identification (RFID), Short domain mobile technology can fulfill the IoT ideal of communication between the physical world and the Internet and Sensor networks. Nowadays, more research is going on wireless network technologies, which have the capability to deliver inconspicuous wire-free communication. Various technology must be provided for the collection of databases about the physical thing (environment) and effective object monitoring. The application and interaction of physical things and the Internet should be accomplished by installing proper software to manage the network's resources, devices, and distributed services. In the case of failure of the system, the software should control and predict self-configuration and auto-recovery. After defining the proper technology and software, hardware selection becomes a difficult task in IoT. The research still is required to develop hardware having radical, a low-energy, multi-purpose framework on a component that will have the ability of self-adaptiveness and self-organization. The dedicated algorithm is used as a component with the chipset of hardware system with an ultra-low power very large scale integrated (VLSI) circuits [1–3].

2.4 LITERATURE REVIEW

When it comes to data-gathering connected to industries and cities, most operations will undergo exponential development in the next years. Sensors that are highly connected and networked will be widely used in a variety of applications. Automation of industrial processes, and, in a larger sense, automation of processes involving all social actors, has resulted in a widespread understanding of tremendous achievements in "cyber-physical systems" and IoT. Cyber-physical systems are the consequence of providing computational and communicational capabilities to physical components. The use of CPS in industrial sectors such as manufacturing, logistics, data, and municipal administration are all possibilities with significant economic, societal, and environmental implications, as these systems may generate pivotal opportunities in fields such as energy interfaces. When it comes to the IoT and Automation, the architecture, engineering, and construction (AEC) sector has generally moved at a slower pace than other more automated and industrialized areas. The rate of acceptance of such technology has been slow. All AEC main players recognize the enormous room for development in areas like sustainability, risk management, clean energy, and infrastructure management. Similarly, one of the most significant potentials for growth in civil engineering is automation innovation.

In some domains, the use of sensors integrated within city or rural infrastructures is well-progressed, while in others, it is still in its infancy. There are numerous examples of physical/chemical sensor integration in water management, energy efficiency, and transportation. Nonetheless, construction process automation or sensor embeddedness in structures is not yet commonplace. The use of sensors within structures is confined to select (though numerous) examples of Structural Health Monitoring that are highly controlled and rigorous. It is fair to say that the vast collection of data from sensors in structures is still in its early stages.

If you're reading this book, it's very likely that you've already heard the terms IoT with a web of things (WoT). Hopefully, you are looking forward to understanding what this trend is all about, or you have already known the importance of this topic and the popularity of same. Maybe you are a civil engineer and realize what IoT could mean for your industry, and you would like to gain the hard technical skills needed to build web-connected products and services.

We have already discussed what IoT means, what new types of applications and scenarios IoT enables, and how this can change the technology and business landscape for the coming years in Construction Industry. This chapter will cover in brief all the web technologies and tools that will help you make the IoT a reality for the construction industry (Figure 2.3). On the other hand, we believe that starting with some background will help you better understand what the IoT really is and how you can use it in your own projects, not just stick to the superficial and stereotypical descriptions of it [2–6].

Two decades ago, it only existed in science-fiction novels and books, where everybody could feel the world through digitalization. They could use sensors, analyze, store, or exchange information on mobile, laptop, or desktop. Today, those scenarios increasingly became smart things. A smart item is a physical object that has been digitally enhanced with one or more of the following features:

- Sensors-simulation;
- Actuators (displays, sound, motors, and so on);
- Computation;
- Communication interfaces.

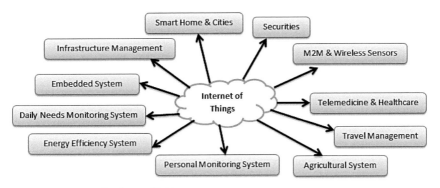

FIGURE 2.3 Applications of IoT.

2.5   IOT IN IT ENABLE THE CONSTRUCTION INDUSTRY

IoT in construction is basically based on utilizing of internet-connected sensors in the job sites or in-situ or may be worn by laborers while they are on duty. For construction, IoT devices can collect certain types of data about the performance of laborers and workers, and activity of the same with circumstances on the construction site, and then send the data to a central dashboard, where it will be examined to aid decision-making.

Occasionally, since the last two decades, the devices with internet connections in them were generally computers and mobile phones. However, now is recent time, with a gadget known as a chip, a wide range of sensors can be quickly changed and inexpensively upgraded (similar to a subscriber identity module (SIM) card). The name "internet of things (IoT)" is used since it is so small and insignificant. It is now possible to connect various gadgets to a central database, from the smartwatches that most people wear nowadays that measure heart rate to temperature sensors and vibration monitors. This means that a tiny gadget or machine can monitor many more features of a place in real-time. This is how the IoT is beneficial and popular these days, with significant consequences [8–10].

2.6   IOT IN CIVIL ENGINEERING

Civil engineering is, without a doubt, the most ancient engineering subject. Even though history claims it dates to the third century. Archimedes, a Greek mathematician who flourished between 287 and 212 B.C., is best known for inventing the screw, which allowed projects like the Pont du Gard, a Roman aqueduct, to be built.

There are numerous courses and curriculum in civil engineering education connected to a wide range of subjects, including structure, geotechnical, mechanics, hydraulics, geosciences, transportation, building materials, construction, survey, and so on.

Following that, these streams are linked to highly professional subjects like structural design, soil mechanics, fluid mechanics, water supply engineering, hydraulic structures, building foundations, bridge engineering, highway, airport, and railway engineering, drinking water and wastewater treatment, structural analysis, and so on. Figure 2.4 illustrates some of the foundational fields of study in which curricula are defined.

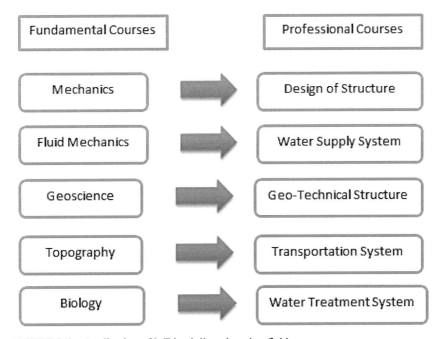

FIGURE 2.4 Application of IoT in civil engineering fields.

Talking about education in civil engineering, there are very few experiences in which the above-mentioned technologies are used that can be found in the literature.

2.7 HOW CAN WE APPLY IOT IN CIVIL ENGINEERING?

The connection between the physical and the virtual worlds has bought new levels of ecosystems into our society. The IoT or ubiquitous sensors used in urban areas are only a fraction of those ecosystems. IoT has intense consequences for new generations, and as a result, in fields like education, trends, and other schools of thought, IoT is emerging, particularly in as robotics and control engineering kingdom. However, thanks to these trends in the construction industry to control are also making a new role. The curricula of civil engineering are infused with commonplace strategies related to automation, physical-to-digital bridges, the systematic use of monitoring devices, and human-computer interfaces. Traditionally, the students of civil engineering in classrooms have described the problems of civil engineering by using classic theory involving a significant amount of mathematical background as well as a thorough understanding of the phenomena. These classic theories are generally been derived from experimental observations from laboratories by civil engineers decades before, establishing various formulae, results, and research. But the development of single or multivariable experiments associated with these fields is still under of the greatest experiences for students at diploma, bachelor, and especially master levels. Similarly, numerical simulations are still increasingly used in education. However, despite of the direct link between the physical and virtual principle in which the curriculum is based, few educational programs in recent times, found in some universities, provide bridges between the existing gap between the physical and virtual world. Therefore, the educators are now starting to face a complex challenge in the field of civil engineering. On the one hand, the continuing trends in monitoring, control, automation, the construction of intelligent cities and the IoT will eventually shape the professional sector of civil in the coming years. However, the study of the use of sensors, electronics, programming, and other technologies that represent the cornerstones of this transformation lack in the curriculum of civil engineering for the students.

Our first application will be dealing with the physical-to-digital interfaces development in real practical problems of civil engineering using mobile phones. These problems are generally reproducible in educational facilities generally found in schools of civil engineering worldwide.

The experiment developed in traditional civil engineering in laboratories include the measurement of torsion, pressure, stress and strain, discharge, water levels, material characterization, soil tests, concrete tests, deflection of beams, aggregate value tests, etc. These magnitudes are measured in the research labs with great precision by means of tested and specialized equipment. Sophisticated user interfaces developed typically by the hardware providers allow gathering and visualizing information in a meaningful way. But there are often some disadvantages in few places; firstly, the cost of these hardware laboratory machineries is too much; secondly, the limitation of facilities in research labs takes place in some newly opened colleges, etc., thus the proper tool of education being not available to the vast majority of students.

However, the costs and accessibility of electronic devices have decreased remarkably in recent years. There is available technology that may smoothen the reproduction of traditional laboratory experiences by a number of students to a reasonable extent, in terms of accuracy and precision.

In some courses, it may be an ideal scenario, the one in which student develops IoT applications easily with correct use of sensors, microcontrollers, and visualization.

> Firstly, array vast in number, of sensors, are both technically and economically available and accessible. Some of the physical magnitudes, such as distance, light, discharge, pressure, temperature, humidity, noise, acceleration, etc., can be measured with the help of these devices. The students of civil engineering are generally acquainted with how these devices transform the variations of physical magnitude in variations related to civil. Thus, the theory and behavior of analog and digital sensors need to be explained theoretically and practically.

> Secondly, the prototyping of electronic boards has become such popular device these days that provide a simulation interface for beginners. The mechanics of the interfaces are quite similar, even if these prototyping boards may not give the same or accurate performance that professional equipment used in laboratories. As civil engineering students have mostly been exposed to practical experiences involving electronics, the basic concepts of electricity and signals (such as analog-to-digital conversion (ADC)) need to be described in class to the civil engineering students beforehand.

The above two points can easily be introduced to IoT in the civil engineering field through their curriculum, thus making it totally applicable in the near future. The main objective of the proposed learning environment is to introduce a closure of the existing physical-to-digital gap among civil engineering students. This environment consists of an immediate experimental experience in which all students are provided with a brief and necessary tool and with a collaborative working space. Expert educators and technicians should be available in the schools for help at particular pre-scheduled hours for clearing any quarry of the students.

2.8   OBSTACLES OF IOT IN THE CONSTRUCTION INDUSTRY

Firstly, an anonymous person or an outsider, someone belonging to a rival company, may find out if a hacker can gain access to a company's IoT database. It may be a goldmine of crucial and authentic data. Criminals may benefit from a list of all your machinery's locations that are now placed or where valuable items are stored. Many people, including labor organizations, may protest on privacy grounds, citing worries about tracking workers' physical movements. Because most IoT devices are inexpensive, many job site owners will have to persuade their client offices that they are worthwhile. When working on smaller locations where all inspections can be completed in a matter of minutes, it may seem pointless to invest in sensors and learn to utilize a dashboard when the work can be completed manually. For the time being, IoT construction solutions are more likely to be deployed on large construction and civil engineering projects. The effectiveness of sensors is determined on how they are used. When the primary risk might not make sense to spend money on moisture sensors. Sensors must be intelligently positioned and chosen, and many construction businesses will face a steep learning curve before reaping the full benefits of these instruments. Placing IoT sensors around a facility will not fix problems on its own.

2.9   POTENTIAL USES OF IOT IN CONSTRUCTION

IoT is already being used in construction companies all around the world. The following are a few examples of how IoT can be used in the construction industry:

1. **Working Conditions That are Safe for Staffs:** Practices such as wear a wristband or use a clip-on device. If every employee on a job site wears a wristband or wears a clip-on device, data on their activity and movements can be utilized to detect any accidents or unsafe behavior. An example of a New York construction firm that provided clip-on IoT devices to workers for their safety. When the device is dropped by three feet or more, the chip-on device sends an automatic notification to the company's site safety manager, or safety department.

2. **Resource Management Improvement:** IoT devices could significantly improve resource management on construction sites, such as how many hours are wasted searching for materials on construction sites, how many liters of fuel are burned by idle engines, and how much time laborers are underemployed when they could be supporting other tasks, etc.

   You can locate all machinery, personnel, and materials promptly if they are all chip-connected to the internet. There are considerable cost and time savings to be had here. Even monitors placed on a truck by an IoT company or a construction company utilizing. IoT solutions can track the location and activities of a wide range of resources.

3. **Better Upkeeping and Reporting:** IoT devices can provide constant input about information and circumstances at both in-situ and under-construction sites when sensors are attached at the construction site. Sensors can detect unusual vibrations on machinery that indicate it needs to be repaired, detect rises in humidity that might alert your inspection teams to a damp issue, and even help prevent fires, as in the case of IoT building construction business pillar.

The area at Plan Radar is exciting. The By delivering app already functions as an IoT device for construction, providing real-time ticketing information for site problems to maintenance employees. However, with advancements in IoT sensors, this might be accomplished even more quickly. Construction companies are already reaping the benefits of the IoT. It is assisting in the following areas:

- Improving the construction industry's health and safety;
- Insurance premiums to be reduced;
- Reduce squandering and steal;
- Enhance more proactive (less expensive) maintenance;
- Increase resource management efficiency.

2.10   FUTURE SCOPE IOT IN IT-BASED CONSTRUCTION

The motive of operations in which the construction business has developed is what is called "the IoT platform." To optimize the process of decision-making and achieve the best resource management, employees, and fleets, the IoT platform aims at providing manufacturers of equipment of construction and end-user contractors with comprehensive analysis and also reports on different elements of their works as well as assets. Sensors are the key in the building field. Sensors provide all the information required to enhance decision-making with the real-time worksite and personnel data to be captured.

The platform offers many extensive details, which may be picked as it may be deemed. This allows for effective and cost-efficient alternatives. Virtually any important information may be obtained, from the consumption of fuel to the production of equipment, from the maintenance services to future forecasts. Moreover, most features are fully adaptable to fulfill special contractor requirements (e.g., alarm bell and notifications, the layout of the dashboard, CAN-Bus parameters, measurement units, languages, etc.).

Data collected, known as big data, offers several analytical methods:

1. **Descriptive:** A thorough account of present conditions and an understanding of what has transpired in a particular work setting;

2. **Predictive:** To ensure reliable product performance predictions and prevent potential malfunctions or injuries from occurring;

3. **Prescriptive:** Know how the process and workflow via projects and plans that feed the machines are optimized and automatized in real-time.

The main advantages of obtaining bigger data are cost savings and a greater understanding of what clients require.

The IoT is one of the most intriguing new advancements in our sector. According to McKinsey, consultants of around $1 trillion on work locations by 2025. This means that it will become a more widespread aspect in the building sector as a whole. The IoT in construction is surely one to watch in recent times, with a wide range of applications and growing benefits.

Smart, networked building equipment can track the way it is utilized. By accumulating previous data, the IoT platform will be able to develop patterns of information to better understand the demands of customers and the environment. In fact, the data acquired from the intelligent goods and systems that are connected give precise information on the usage, utilization, and ignoring of product features.

IoT in the IT-based construction industry has the potential effect of reducing all kinds of accidents that might take place at work sites using digital solutions, like sensors on workers, their movement and position can be tracked by the same, clothing, and hardhats used, wearable of the staffs that monitor for hazardous materials present in the construction site, sensors that will monitor and alert when workers enter unsafe areas in the site, etc. The list goes on for ways in which digitalization them and all means. IoT can help keep projects moving forward, without making delay of completion of project time.

In the world, IoT, their technologies, digitalization, and data are used to improve their activities by construction businesses. Australia is seeing a rising tendency towards partnerships with technology suppliers in constructing, engineering, and consulting organizations, which are developing mutually intelligent solutions. The initiative offers advantages such as tracking of the flowing of assets and goods in multi-stored buildings in real-time, monitoring the total number of persons present on site, and providing firms for construction with data for logistics optimization, theft, inventory maintenance, and safety risk mitigation. Building may flourish with data and digital tools. About 95% of the construction organizations now have informed KPMG (audit, tax, and advisory services are provided by a global network of professional businesses.) that IoT included in the industry are changing their industries fundamentally, while PwC says that 98% of construction organizations anticipate digital instruments to improve their work efficiency by about 12%. IoT instruments is allowing building enterprises to make use of real-time data for assisting them flourish in such industry which is famous for the overruns of price, expensive equipment, deficiencies of skilled work and increasing short project schedules.

IoT has now been integrated for smart buildings as BIoT (building internet of things) focusing on more sustainable buildings focusing on multifunctional management console. Smart building aims to create buildings with more comfort and lesser expenditure with minimum adversity to the environment. The benefits of building automation systems go beyond energy saving. The development of intelligent buildings will obviously be driven by intelligent cities and their intelligent buildings will maximize the experience of the residents. Implementing the climate change policy will need building efficiency.

IoT enables to remotely sense objects and control them across existing network infrastructures, thereby providing more opportunity for the physical world to be integrated directly into computer-based systems and improving, in addition to decrease human intervention, efficiency, accuracy, and economic benefits. Increasing IoTs via the use of sensors and actuators makes it an example of the most generic class of cyber-physical systems, technologies such as intelligent networks, virtual power stations, using clever households, smart transit, and smart cities.

As already discussed, IoT has immense benefits in the construction industry, and there are many more to be identified in the coming years. Construction is the type of industry that cannot thrive without the human grind. Hence it becomes the responsibility of the construction company or the organization to ensure the health and safety of its workers and simultaneously be mindful of the profit and reduce waste for better sustainability. IoT can help in ensuring all of these for the organization with minimum effort and toil.

As IoT seems to be more omnipresent, the way in which the building sector changes has a bigger effect. IoT enables each stakeholder to understand what is occurring in every phase of the building process in real-time, from design to the actual building, post-construction, and how the facility is managed throughout service. As the building business evolves at a glacial pace, construction businesses using technology to handle common workplace issues successfully and simplify procedures profit from higher efficiency and enhanced response to the growing needs of the sector.

IoT in construction has come a long way since its introduction to the industry, yet there are certain shortcomings that are yet to be met. Although it has already been established that IoT can support the construction industry in the best possible way and in the coming days, this could find more ways for improvisations in the industry.

KEYWORDS

- **artificial intelligence**
- **building information modeling**
- **computing-assisted design**
- **Internet of Things**
- **unmanned aerial vehicles**
- **virtual reality**

REFERENCES

1. Rupamkumar, V. M., Jayeshkumar, R. P., & Amitkumar, D. R., (2020). Feasibility Study of Internet of Things (IoT) In Construction Industry. Journal of Emerging Technologies and Innovative Research, 7(5), 447–452.
2. Ram, B. M., (2021). *IoT implementation in Construction Industry (Internet of Things), 12*(1).
3. Yaser, G., Majid, A. A., Ismail, A. R., & Muhammad, M. A., (2019). "Internet of things in construction industry revolution 4.0," January 2020, *Journal of Engineering Design and Technology,* DOI:10.1108/JEDT-06-2019-0164.
4. Sachin, K., Prayag, T., & Mikhail, Z., (2019). Internet of Things is a revolutionary approach for future technology enhancement: a review. *Journal of Big Data, 6*(1), DOI:10.1186/s40537-019-0268-2.
5. IoT Applications in Construction. https://www.iotforall.com/ (accessed on 11 October 2022).
6. https://www.smartdatacollective.com/5-incredible-iot-applications-in-civil-engineering/ (accessed on 11 October 2022).
7. Viren, C., & Arbaz, K., (2017). *Application of Internet of Things in Civil Engineering Construction Projects-a State of the Art.*
8. https://engineeringcivil.org/articles/civil-engineering-the-internet-of-things/ (accessed on 11 October 2022).
9. Arka, G., David, J. E., & Reza, H. M., (2020). Patterns and trends in Internet of Things (IoT) research: future applications in the construction industry. *Engineering, Construction and Architectural Management,* ISSN: 0969-9988, 2020.
10. Ram, B. M., (2021). "Cost-Effective Voice-Controlled Real-Time Smart Informative Interface Design With Google Assistant Technology." *Machine Learning Techniques and Analytics for Cloud Security* (Chapter 4, pp. 61–79), https://doi.org/10.1002/9781119764113.ch4, 2021.
11. Rolando, C., Hector, P., Álvaro, T., & Maria De, G. E. C., (2018). "Development of IoT applications in civil engineering classrooms using mobile devices." *Computer*

Applications in Engineering Education (Volume 26, pp. 1769–1781), https://doi.org/10.1002/cae.21985, 2018.

12. Radu F. Babiceanu & Seker, R., (2016). Big data and virtualization for manufacturing cyber-physical systems: A survey of the current status and future outlook. *Computers in Industry, 81*, 128–137.

13. Dave, D., Kubler, S., Främling, K., & Koskela, L., (2016). Opportunities for enhanced lean construction management using internet of things standards. *Automation in Construction, 61*, 86–97.

14. Bibri, S., (2017). The IoT for smart sustainable cities of the future: An analytical framework for sensor-based big data applications for environmental sustainability. *Sustainable Cities and Society, 38*, 230–253.

15. Ronald, C. Y. L., & Alvin, J., (2018). *IoT Application in Construction and Civil Engineering Works.* IEEE Xplore.

16. Lizong, Z., Anthony, S. A., & Hongnian, Y., (2011). Knowledge management application of internet of things in construction waste logistics with RFID technology. In: *5th IEEE International Conference on Advanced Computing & Communication Technologies* (Vol. 3, pp. 220–225).

17. Harish, G. R., & Venkatesh, K., (2019). Study on implementing smart construction with various applications using internet of things techniques. *International Journal of Recent Technology and Engineering (IJRTE)* (Vol. 7, No. 6C2). ISSN: 2277-3878.

CHAPTER 3

Object Detection from Video Sequence and Its Analysis for Application in Wildlife

SHRUTI SHOVAN DAS[1] and SOUMYAJIT PAL[2]

[1]*Undergraduate Student, Asutosh College under University of Calcutta, Kolkata, West Bengal, India*

[2]*Faculty of Information Technology, St. Xavier's University, Kolkata, West Bengal, India*

ABSTRACT

Object detection is a key feature needed by the majority of computer and robot vision systems. The most recent study in this field has made significant strides. During the last few years, there has been a successful increase in computer vision research. Part of this success may be attributed to the adoption and adaptation of machine learning methods, while others can be attributed to the invention of novel representations and models for specific computer vision challenges or the development of efficient solutions. Object detection is one area that has made significant improvement. As a consequence of the enhanced computational capacity of current processors, progress is being made on a daily basis. More advances like this will boost the chances of Object detection and other related operations. Object detection, given a collection of object classes, consists of determining the location and size of all object instances, if any, that are present in a picture or an image in a series (e.g., video). Thus, the goal of an object detector is to locate all object instances of one or more provided object classes

System Design Using the Internet of Things with Deep Learning Applications.
Arpan Deyasi, Angsuman Sarkar, and Soumen Santra (Eds)

independent of scale, position, posture, camera view, partial occlusions, or lighting conditions. Object detection is often the initial stage in many computer vision systems since it allows additional information about the identified object and the scene to be obtained. Once an object instance (e.g., a face, body, thing, etc.) has been detected, it is possible to obtain additional information, such as: (i) recognizing the specific instance; (ii) tracking the object over an image sequence (e.g., tracking an object in a video); and (iii) extracting additional information about the object (e.g., determining the subject's gender).

It is also possible to: (i) infer the existence or placement of other items in the picture (e.g., an obstacle may be near a face and at a comparable scale); and (ii) better estimate other information about the scene (e.g., the type of setting, etc.).

In this current manuscript, we discuss how the detection of a single object class helps us in detecting and reducing the amount of video analysis required to record and detect objects (e.g., animals or birds, etc.) appearing in front of the recording device. The single-object detection approach is preferred here as, if the algorithm is able to detect a single object precisely, then it may be able to detect multiple objects in front of it as well. Since the graph of the duration of stay of the object is plotted with respect to the single object detected hence it can be inferred if the algorithm can detect a single object from a collection of objects, then the graph will also be plotted for it accordingly. Thus, a reduction of effort is required to analyze video recordings of great length.

Object detection has been utilized in a wide range of applications. Each of these applications has various requirements, such as processing time, occlusion resilience, rotation invariance (e.g., in-plane rotations), and detection under posture changes. While many applications include detecting a single object class (e.g., faces) and from a single perspective (e.g., frontal faces), others require detecting numerous object classes (vehicles, birds, etc.), or a single class from various viewpoints (e.g., side and frontal view of vehicles). Most systems, in general, can detect just a single object class given a limited number of perspectives and positions.

3.1 INTRODUCTION

Vision is an important aspect that is required by computer and robotic systems. Object detection plays an important part in achieving this. The pace of research and development has been high in the field of research and development in computer vision in recent years. And, as a result, the field of object detection has made progress as well.

The successful growth of computer vision research has occurred due to incorporating machine learning techniques and adjusting accordingly. The increased computing power of modern-day processors has played a huge part in the contribution of the development of this field. It is projected that such developments will lead to the growth of prospects of object detection and other allied activities [1].

Object detection has vital importance and many other related activities or processes are dependent on it. This paragraph will shed light briefly on some of them. Detecting the position and span of all instances of objects that are there in an image sequence whereby there exists a collection of similar object classes is also contained under object detection. Irrespective of size, placement, posture, and scene with respect to a camera device, partial occlusions, and lighting conditions, the task of an object detector is to recognize the presence of one or multiple object classes. Perception of a given instance, tracking of the same over the video, information extra to the object (e.g., gender), the existence of other objects in the scene, evaluation of the scene, etc., are also obtained following the detection of the object [1].

In this chapter, we go through how object detection can help us record and detect items (like animals or birds, etc.) that appear in front of the recording device while also minimizing the amount of video analysis needed to do so. Here, the single-object detection strategy is recommended because, if the algorithm can accurately identify a single thing, it might also be able to identify several objects in front of it. Since the graph of the item's duration of stay is presented in relation to the single object discovered, it can be deduced that if the algorithm is able to identify a single object among a group of objects, the graph will likewise be plotted for it in the appropriate manner. If it can detect one object from a collection of objects in front of the lens of a camera, then during video analysis for editing purposes etc. by professionals the cost of analysis and editing reduces as we only focus on extracting and editing only those sections of the recordings that have detected objects successfully in front of the camera within the time span of deployment of it. Hence, this acts as a cost-effective measure and improves the productivity of professionals working with wildlife recordings.

Object detection finds usage acceptance over a range of applications. Time required for processing, ability to be robust to occlusions, rotational constancy, and detection under posture transitions are some of the factors upon which the requirements of these applications depend upon. The different types of object detection that applications take into consideration are single object class detection from a single, detection of several object classes, or detection ability of a single class from more than one view. Predominantly, most systems could detect a single object class from a limited collection of similar views and postures [1].

3.2   EXPERIMENTAL OBSERVATIONS AND PROPOSALS

3.2.1   OBSERVATIONS

The amount of lighting available at a point in time affects the intensities uniformly all over the image with respect to time (i.e., the overall image experiences lightness or darkness due to lighting conditions). In a real-life scenario, the camera experiences changing lighting patterns with respect to time. Hence, we encounter problems as the proper adjustment of application doesn't take place according to changing lighting conditions.

Simulation is conducted using an external webcam attached to a laptop by varying the lighting conditions. Once the application starts running, the video of the scene is captured. When an object appears before the webcam, the detection process begins. Initial background subtraction is followed by binary thresholding to make the background pixels black and the foreground pixels white. The image is then dilated to increase the size of the detected object, which is denoted by a rectangular region enclosing it. The timestamps of objects appearing and leaving the scene are recorded and plotted using a motion graph. This gives an idea of the duration for which the detected object was in sight.

Here, in Figures 3.1–3.4, we can see how lighting conditions affect the application that is supposed to record video for long hours. Hence, if proper calibration or adjustment of application does not take place, it shall become a problem while in use. Here, the change of lighting conditions in Figures 3.1–3.4 simulates the day-night and night-day transitions.

FIGURE 3.1 The application is started at night time or with poor lighting. It is still able to detect the face.

FIGURE 3.2 As soon as the lighting condition improves, the object detection algorithm detects the whole window as an object.

FIGURE 3.3 The application is restarted with good lighting conditions. It is able to detect the face.

FIGURE 3.4   The object detection window detects the whole object detection window as an object is detected.

3.2.1.1   IMPROVEMENT DEDUCED

If the application is restarted after a suitable fixed interval, it shall help the application to adjust to its corresponding lighting conditions, and no other major change is required as well to the application. Ex-Let the application restart after 1-hour intervals, and hence it will adjust itself automatically with lighting conditions due to cloud factors, change in position due to the Sun, Moon, or any other light source. Each time the application restarts, it starts to record a new video file for that duration. Assume a bright light source appears in front of the lens of the camera.

We can see in Figures 3.5–3.8 that as the light source is emitting light energy, the rays of light surrounding the source are also detected as an object by the object detection algorithm. Thus, in this way, the object appears to be detected by a detection window of larger size than the total area of the light-emitting region.

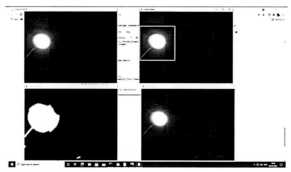

FIGURE 3.5   Bright light source emitting light energy appears in dark lighting conditions.

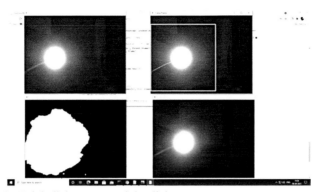

FIGURE 3.6 Bright light source emitting light energy moves.

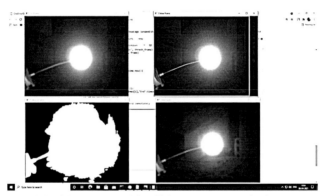

FIGURE 3.7 Bright light source emitting light energy moves further.

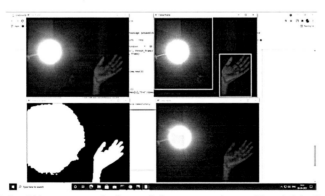

FIGURE 3.8 Object appears in front of the camera along with a bright light source emitting light energy.

In Figure 3.8, we can see that the application was able to detect another object (here, the hand). We can propose in the next section that if it is possible to detect light sources or other objects (trees, rocks, bushes, etc.) that are not desired for wildlife object detection, we may label such undesired objects as background and consider the rest detected objects as foreground (desired objects to be detected) as mentioned in the proposal section of this chapter.

Similar observations are also noticed for light-emitting sources present in good lighting conditions in Figures 3.9–3.11.

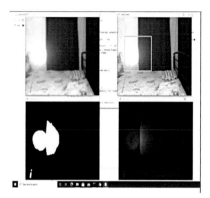

FIGURE 3.9   Light source emits light energy in well-lit lighting conditions.

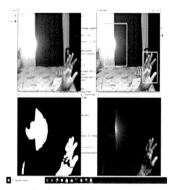

FIGURE 3.10   Object appears in front of the camera lens along with the bright light source emitting light energy.

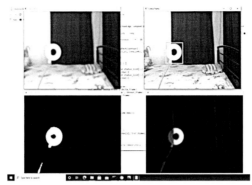

FIGURE 3.11 When the light source is not emitting light energy but is visible due to well-lit lighting conditions.

Here also, in Figures 3.9–3.11, we find the light source emitting light energy as rays have been detected by a detection window of much larger size than the light emitting region of the light source.

We may also apply a similar proposal to classify background and foreground objects like the previous cases, Figures 3.5–3.8, for cases of Figures 3.9 and 3.10 as well.

We see here in this figure that the light source is detected by a far smaller object detected window than in the cases where it was emitting light energy. This is the case when the light source is an artificial light source.

We are also not considering the visibility of an artificial light source when it's not emitting light energy or is turned off in dark conditions, as it would be impossible to see it.

Figures 3.5–3.11 are also simulations of the application with respect to visibility scenarios and object detection scenarios in day or night time. For Figures 3.5–3.10, the light source may be natural or artificial, but for Figure 3.11, it must always be an artificial source in a well-light surrounding environment.

The light sources here may also be identified as classes and classified as background or foreground. Otherwise, if a light source is always present in front of the lens, then a graph may be plotted for the whole duration the light source is viewed as an object present in front of the lens. This will in turn make our purpose of application not fully utilized to reduce our editing or analysis efforts (Figures 3.12 and 3.13).

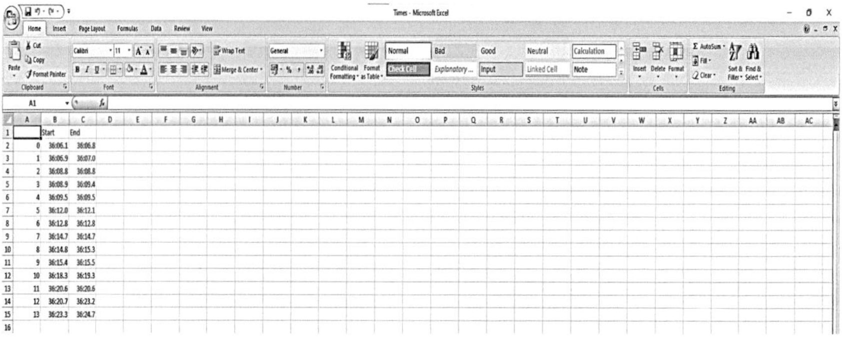

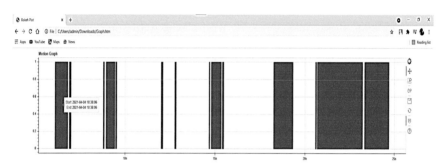

FIGURE 3.12   CSV file storing details of the object detected with its 'start timestamp' and 'end timestamp' to show the durations of objects detected.

FIGURE 3.13   Motion graph showing durations of the object detected with their 'start timestamp' and 'end timestamp' as we hover over the graph.

Hence, these are all the observations that we note from the application. Here, objects refer to 'wildlife' in objects detected.

3.2.2 PROPOSALS

The following is the list of proposals that have not been implemented in the source code the observations have been studied from but may be applied for future improvements of the application:

- Objects may be identified with respect to well-defined classes based on available surroundings of the placement of the lens and hence be checked if it resembles any class from the already defined set of classes. After identification of such classes, they may be classified as background or foreground so that when an undesired object is detected, like a tree, stone, or light sources like the sun or moon or an artificial light source, they can be classified as the background so that the graph doesn't start to plot with respect to the undesired objects. This will be helpful if a tree is always present in front of the lens and the object detection algorithm starts to identify it as the desired object and hence draw a graph for the whole duration it was present (here, always), thus leading our intention of using the application for our desired purpose not get satisfied.

- Deep learning and machine learning techniques can be utilized effectively in identifying classes and hence making object detection smoother and more efficient.

- Contextual information and temporal features: Incorporating background details could help to increase the qualities of being efficient and robust for the application. However, it is still a dilemma as to which moment (i.e., prior or later to object detection) and how these details could be integrated. The solution is yet not known for this, but it is assumed to possess a verifiable solution. There have been ideas for a few solutions, like the addition of the usage of spatiotemporal context [2], spatial structure amongst visual words [3], and semantic information focusing on plotting semantically associated characteristics to visual words [4], amongst others [5–9]. There can be advantages to Temporal features [10, 11] as a good number of approaches assess the detection of objects in a single frame [1].

- High resolution of apparatus is required for the best results.

- To increase the efficiency of the analysis of the huge amount of data, we can use GPUs.

- For cameras overhead (e.g., in drones or overhead cameras) during the time of dawn or dusk, we get shadows beside objects like in Figures 3.14 and 3.15.

We get such shadows of objects (animals, trees, etc.) for certain times of the day, like dawn and dusk, when light appears at a certain angle, and we have overhead cameras.

FIGURE 3.14   The shadow of a rhinoceros (toy).

FIGURE 3.15   The shadow of a bull (toy).

If we use techniques such as machine learning and deep learning to identify the such pattern of objects, then we can easily identify objects accordingly and also classify them as what they are. This becomes very much useful for drone shots that are being used these days extensively for filming and editing.

- For objects that are heavily camouflaged, it is very difficult to detect them. Ex-Snow leopards, chameleons, etc.

  To detect them, we may use trace the edges of the surrounding environment initially and set them fixed. Next, the comparison is made with the initially traced edges of the surrounding environment and the edges that are being traced with respect to time continuously. There may possibly be an intrusion by an object if a disruption in the continuity of the edges takes place.

  Example situation – let it be a snowy mountainous region inside a snow leopard habitat. Let the camera be set up beside a water body, with the lens facing the other side of the bank. Initially, the edges of the bank are traced and saved. Now as with each passing moment, the edges are traced, and compared with the initially traced edges, we find that no disturbance or intrusion has taken place. Until the snow leopard appears, which is highly camouflaged with the surroundings, and starts to drink with its tongue. As soon as the tongue breaks the continuity of the edge of the bank, the algorithm detects an intrusion and suspects that there is an object present at that particular position. The graph starts to be plotted accordingly for the duration the snow leopard drinks water and hence in this way highly camouflaged objects could also be detected.

- A timer is maintained with a specific interval set, after which the application restarts to adjust to its surrounding lighting features, etc. Each time the application restarts, a new video recording file is created and starts to record until the next restart.

- If all or majority of the pixels of the image with respect to time (i.e., video sequence) get affected within a very short duration of time, then we can make a case that the object is too close to the lens of the camera (Figure 3.16).

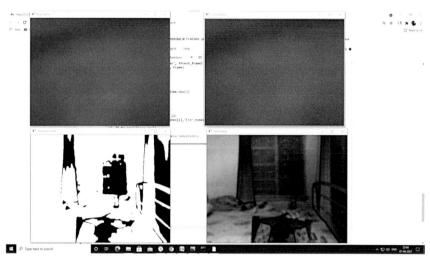

FIGURE 3.16   Object too close to the lens of the camera.

- Sometimes, we may also find object detection windows randomly detecting objects that do not exist. Such cases detect objects for a very short duration of time. We can specify a minimum duration of the object detected, after which the graph to be plotted should start. In this way, we can get rid of such random object-detected windows that pop up during the runtime of the application by setting a threshold time duration.

3.3   DISCUSSIONS

3.3.1   CURRENT RESEARCH DIRECTIONS

1. **Multiple-Class:** There are many examples of the usage of programs where detecting multiple object classes is needed. The types of classes that the system can support and manage without accuracy loss and the rate of processing are essential factors in this case. In Refs. [12–17], there are works that have dealt with multi-class detection issues. The same form of representation for numerous object classes, and multi-class classifier development that is precisely designed to detect multiple classes have managed to deal with the efficiency issue. One of the rarely available works

for multi-class object detection is discussed in Ref. [18]. Around 1,00,000 object classes were taken into consideration for this. In this application, if the algorithm was developed to be able to detect more than one object class, then it would be much more advantageous with respect to future use as the invention or creation of new objects, fashion changes, etc., are a frequent occurrence in the 21st century, and hence our object detection systems will require to be updated continually to add new classes or updating already existing classes [1].

2. **Multiple-View, Multiple-Pose, Multiple-Resolution:** Amongst the approaches that are currently in usage, a majority of such detect a single object class with respect to a single view. Thus, they lack the ability to manage multiple views or significant posture changes. This includes an exception of deformable part-based models that are able to handle a few posture changes. Learning subclasses [19] or by taking into account postures/views as separate classes [16] are the two ways some works have ventured to detect objects. As a result, there have been significant improvements in the fields of efficiency and being robust. Furthermore, there also has been the development of multi-pose models [20] and multi-resolution models [21]. With the ability to detect multiple views, multiple poses, or multiple resolutions would make the application highly robust and dynamic to be used in real-time as in wildlife we may find animals in numerous poses and views in front of the lens of the camera [1].

3. **Efficiency and Computational Power:** One of the most important variables that must be considered is efficiency in any object-detection system. When efficiency is an essential requirement the coarse-to-fine classifier is generally the first kind of classifier to be taken into account [10]. Methods like reducing the total amount of picture patches classification needs to be performed [22] and efficiently identifying many classes [16] have also been deployed. Real-time performance is not implied by efficiency, and works such as in Ref. [23] are robust, scalable, and efficient and not adequate for real-time problems. But, some approaches (for example, deep learning) can be run in real-time utilizing specialized hardware (e.g., GPU). GPU development has been progressing at a good pace

during recent years, and more development is expected in this field in the upcoming years as well with much greater abilities embedded in them. GPUs are also an important part of gaming in these years, and a large number of people own GPUs these days [1].

4. **Occlusions, Deformable Objects and Interlaced Object, and Background:** Although relevant research has been [24] done there exists no compelling solution to deal with partial occlusions. Objects where objects and background pixels can be found not interlaced are known as "closed" objects. Detecting objects that are not "closed" is still a tough problem that exists. Examples of such problems are hand detection [25] and pedestrian detection [26]. More development on the deformable part-based model is still necessary [23] as it has been able to achieve success only to some extent in this type of problem. Provisions to handle occlusions, deformable objects, and interlaced objects and background shall be highly useful for further development of the application as such situations may occur during use in real-life environments, especially with wildlife that is highly camouflaged in their surroundings [1] (Figures 3.17 and 3.18).

FIGURE 3.17   Occlusions as cage bars.

FIGURE 3.18 Occlusions as one bird are in front of another and are also in close proximity.

This may confuse object-detection algorithms to identify them as different objects or as a single object. Objects may also use their color as camouflage with surroundings for cases of interlaced objects and backgrounds.

5. **Contextual Information and Temporal Features:** Incorporating background information (e.g., about the scene and the presence of other objects) could improve the efficiency and robustness of the application. But, the procedure of 'when and in which way' to do this, that is before or after the detection of objects takes place is still a problem that exists in the open world. Some solutions have been proposed, and those among many others [5–9] are the use of (i) spatiotemporal context [2]; (ii) spatial structure among visual words [3]; and (iii) semantic information attempting to map semantically relevant aspects to visual words [4]. Because most algorithms examine object detection in a single frame [1], temporal characteristics might be advantageous [10, 11].

This is another very important feature that shall be necessary for the usage of such applications in real-life scenarios, as cameras can then have more information regarding their surroundings. This will, in turn, help them identify objects that are to be considered background (undesired) or foreground (desired). Thus, making object detection in dynamic situations efficient and effective.

3.3.2   SINGLE OBJECT DETECTION USED IN THE APPLICATION

Single object detection: The focus of the application described in this manuscript is to identify single objects effectively as our graph will be plotted as soon as an object is detected by the object-detection algorithm. The duration in which the object appears is shown by the graph. And during video analysis or editing, we only work with those portions of the video in which the object appeared. Thus, if we have a lioness walking with her cubs or a herd of zebras in front of the lens, if the algorithm of the application can identify one object effectively amongst them, then we can work that duration of the clip for analysis or editing as its graph gets plotted accordingly with respect to the single object detected. This reduces the complexity of developing complex algorithms to identify multiple classes of objects, and we haven't used such complex code for the purpose of working with video clips that are only required when an item emerges in front of the camera's lens.

3.3.3   FUTURE RESEARCH DIRECTIONS

1. **Open-World Learning and Active Vision:** A vital problem is to gradually learn to be able to detect new classes or to gradually learn to be able to distinguish between subclasses after the "main class" has been learned. We would be able to create new classifiers from existing ones without much effort if this could be done in an unsupervised way. Thus, lowering the efforts required to learn new object classes. Creation or invention of new objects, fashion changes/styles, etc., are coming up constantly in society in this century, and hence our systems will also need to keep pace with them. There needs to be always updates, the addition

of new classes, or updates of existing classes. These problems have been addressed with methods based on deep learning and transfer learning methods [27–29] in recent times. This open-world learning is of particular importance in robot applications' cases where active vision mechanisms can aid in the detection and learning [30, 31]. The "open-world learning and active vision" sector is expected to develop in the future as object detection, and computer vision is being incorporated into appliances of daily lives these days at a pace much greater than in the last decade [1].

2. **Object-Part Relation:** There still exists a dilemma during the detection process whether the object should be detected first or the parts of the object should be detected first. And, no clear answer exists. Perhaps, these two processes should be continued simultaneously while both give feedback to each other. How this can be achieved is yet an open problem and is expected to be related with how to use the context information. Also, an object's part may be decomposed into subparts. An interaction amidst several hierarchies emerges, and it is still not clear about what should be done first in general. The dilemma of Object-Part relation also comes into the picture for the application here as if the object detection algorithm were to identify an animal according to their physical features like the horns of cow, buffalo, or deer. Also, if the object were heavily camouflaged in its surroundings with only one such physical feature visible clearly out of camouflage (here, horns) and hence the object detection algorithm searches for resemblance with such classes that it already has prior knowledge about [1].

3. **Pixel-Level Detection (Segmentation) and Identification Background Objects:** We could be interested in recognizing items that are often ignored as background. Rivers, fences, and mountains are examples of "background objects." Most of the methodologies presented in this chapter do not address the identification of background objects. The image is first segmented, and then each segment of the image is labeled [32]. This is how similar problems have been solved in the past. To completely comprehend a scene and effectively detect all items inside it, we will need pixel-level object detection as well as a 3D representation of the scene. Hence,

there exists a possibility that the object detection and image segmentation methods may require to be integrated together. The automatic comprehension of the world is still far to be achieved. And, to achieve this, active vision mechanisms may be required [9, 33]. Objects being identified or classified as background objects can simplify the need to detect desired objects or the foreground. This makes the application familiar with its surrounding environment during its usage and also detects objects that are desired to be detected. This feature can become much helpful for the deployment of such applications or cameras with object-detecting abilities as the surrounding environment is changing rapidly in recent times, and humans are also creating new objects, fashion changes, etc. Embedding these features with the Object detection application will improve productivity, produce desired results, and also improve the robustness of the application with respect to the environment it has been positioned or placed or on [1].

4. **Presence of Adversarial Patches:** There may exist an adversarial patch that can be used to bypass the object detection algorithm toward invisibility [34]. Researchers at KU Leuven generated an adversarial printout patch that can fool an object detection algorithm YOLOv2 [35]. For this to work, the patch has to be held at a specific position and angle. Hence, if the adversarial patches are changed dynamically on a target object, then it can fool the AI much more effectively. It can only trick one particular algorithm. However, it may be possible to use this technique to produce clothing that works like an invisibility cloak for surveillance cameras or other cameras with object detection ability. It can be possible in the future those adversarial patches may exist for object detection algorithms that bypass object detection algorithms towards invisibility.

3.4 CONCLUSION

Human-level efficiency is yet to be achieved in terms of the technology available for object detection. Quad-copters, drones, and service robots in the future, etc., are going to be deployed gradually. Previously mentioned gadgets and appliances are examples of mobile robots and autonomous machines. The need for more object detection systems is being felt as time and technology progress. Several technologies present are a part of appliances used by consumers currently. Such examples can be mentioned as auto-focus in smartphones and assisted driving technologies. Immense progress has been achieved and observed in recent years owing to the technological advancements and efforts of the scientific research community [1].

The current manuscript discusses how the single object detection method may be used to tackle problems in analysis and editing purposes for wildlife documentaries or wildlife filming, etc. The approach can hugely reduce the amount of effort required to go through clips recorded in a large time span of the day the camera records at a fixed position. It increases productivity and reduces the cost of analysis of long video clips manually by professionals for wildlife filming, editing, and analysis. The single object detection algorithm serves the purpose if the desired object is detected. We shall continue to integrate object detection algorithms in our lives as more progress is expected in the fields of object detection and computer vision in the times ahead.

KEYWORDS

- **adversarial patches**
- **contextual information and temporal features**
- **deep learning**
- **machine learning**
- **multiple-pose**
- **spatiotemporal context**

REFERENCES

1. Verschae, R., & Ruiz-del-Solar, J., (2015). Object detection: Current and future directions. *Front. Robot. AI, 2*, 29. doi: 10.3389/frobt.2015.00029.
2. Palma-Amestoy, R., Ruiz-Del, S. J., Yanez, J. M., & Guerrero, P., (2010). Spatiotemporal context integration in robot vision. *Int. J. Human. Robot., 7*, 357–377. doi: 10.1142/S0219843610002192.

3. Wu, L., Hu, Y., Li, M., Yu, N., & Hua, X. S., (2009). Scale-invariant visual language modeling for object categorization. *IEEE Trans. Multimedia, 11*, 286–294. doi: 10.1109/TMM.2008.2009692.

4. Wu, L., Hoi, S., & Yu, N., (2010). Semantics-preserving bag-of-words models and applications. *IEEE Trans. Image Process, 19*, 1908–1920. doi: 10.1109/TIP.2010.2045169.

5. Torralba, A., & Sinha, P., (2001). Statistical context priming for object detection. In: *Computer Vision, 2001: ICCV 2001; Proceedings Eighth IEEE International Conference* (Vol. 1, pp. 763–770). Vancouver, BC: IEEE. doi: 10.1109/ICCV.2001.937604.

6. Divvala, S., Hoiem, D., Hays, J., Efros, A., & Hebert, M., (2009). An empirical study of context in object detection. In: *Computer Vision and Pattern Recognition, 2009. CVPR 2009. IEEE Conference* (pp. 1271–1278). Miami, FL: IEE). doi: 10.1109/CVPR.2009.5206532.

7. Sun, M., Bao, S., & Savarese, S., (2012). Object detection using geometrical context feedback. *Int. J. Comput. Vis., 100*, 154–169. doi: 10.1007/s11263-012-0547-2.

8. Mottaghi, R., Chen, X., Liu, X., Cho, N. G., Lee, S. W., Fidler, S., et al., (2014). The role of context for object detection and semantic segmentation in the wild. In: *Computer Vision and Pattern Recognition (CVPR), 2014 IEEE Conference* (pp. 891–898). Columbus, OH: IEEE. doi: 10.1109/CVPR.2014.119.

9. Cadena, C., Dick, A., & Reid, I., (2015). A fast, modular scene understanding system using context-aware object detection. In: *Robotics and Automation (ICRA), 2015 IEEE International Conference*. Seattle, WA.

10. Viola, P., Jones, M., & Snow, D., (2005). Detecting pedestrians using patterns of motion and appearance. *Int. J. Comput. Vis., 63*, 153–161. doi: 10.1007/s11263-005-6644-8.

11. Dalal, N., Triggs, B., & Schmid, C., (2006). Human detection using oriented histograms of flow and appearance. In: Leonardis, A., Bischof, H., & Pinz, A., (eds.), *Computer Vision ECCV 2006: Lecture Notes in Computer Science* (Vol. 3952, pp. 428–441). Berlin: Springer.

12. Torralba, A., Murphy, K. P., & Freeman, W. T., (2007). Sharing visual features for multi-class and Multiview object detection. *IEEE Trans. Pattern Anal. Mach. Intell., 29*, 854–869. doi: 10.1109/TPAMI.2007.1055.

13. Razavi, N., Gall, J., & Van, G. L., (2011). Scalable multi-class object detection. In: *Computer Vision and Pattern Recognition (CVPR), 2011 IEEE Conference* (pp. 1505–1512). Providence, RI: IEEE. doi: 10.1109/CVPR.2011.5995441.

14. Benbouzid, D., Busa-Fekete, R., & Kegl, B., (2012). Fast classification using sparse decision DAGs. In: Langford, J., & Pineau, J., (eds.), *Proceedings of the 29th International Conference on Machine Learning (ICML-12), ICML '12* (pp. 951–958). New York, NY: Omni press.

15. Song, H. O., Zickler, S., Althoff, T., Girshick, R., Fritz, M., Geyer, C., et al., (2012). Sparselet models for efficient multi-class object detection. In: *Computer Vision-ECCV 2012* (pp. 802–815). Florence: Springer.

16. Verschae, R., & Ruiz-del-Solar, J., (2012). TCAS: A multi-class object detector for robot and computer vision applications. In: Bebis, G., Boyle, R., Parvin, B., Koracin, D., Fowlkes, C., Wang, S., et al., (eds.), *Advances in Visual Computing: Lecture Notes in Computer Science* (Vol. 7431, pp. 632–641). Berlin: Springer.

17. Erhan, D., Szegedy, C., Toshev, A., & Anguelov, D., (2014). Scalable object detection using deep neural networks. In: *Computer Vision and Pattern Recognition (CVPR), 2014 IEEE Conference* (pp. 2155–2162). Columbus, OH: IEEE. doi: 10.1109/CVPR.2014.276.

18. Dean, T., Ruzon, M., Segal, M., Shlens, J., Vijayanarasimhan, S., Yagnik, J., et al., (2013). Fast, accurate detection of 100,000 object classes on a single machine. In: *Computer Vision and Pattern Recognition (CVPR), 2013 IEEE Conference* (pp. 1814–1821). Washington, DC: IEEE.

19. Wu, B., & Nevatia, R., (2007). Cluster boosted tree classifier for multi-view, multi-pose object detection. In: *ICCV* (pp. 1–8). Rio de Janeiro: IEEE.

20. Erol, A., Bebis, G., Nicolescu, M., Boyle, R. D., & Twombly, X., (2007). Vision-based hand pose estimation: A review. *Comput. Vis. Image Underst., 108*, 52–73. Special issue on vision for human-computer interaction. doi: 10.1016/j.cviu.2006.10.012.

21. Park, D., Ramanan, D., & Fowlkes, C., (2010). Multiresolution models for object detection. In: Daniilidis, K., Maragos, P., & Paragios, N., (eds.), *Computer Vision ECCV 2010: Lecture Notes in Computer Science* (Vol. 6314, pp. 241–254). Berlin: Springer.

22. Lampert, C. H., Blaschko, M., & Hofmann, T., (2009). Efficient subwindow search: A branch and bound framework for object localization. *IEEE Trans. Pattern Anal. Mach. Intell., 31*, 2129–2142. doi: 10.1109/TPAMI.2009.144.

23. Felzenszwalb, P., Girshick, R., McAllester, D., & Ramanan, D., (2010b). Object detection with discriminatively trained part-based models. *IEEE Trans. Pattern Anal. Mach. Intell., 32*, 1627–1645. doi: 10.1109/TPAMI.2009.167.

24. Wu, B., & Nevatia, R., (2005). Detection of multiple, partially occluded humans in a single image by Bayesian combination of edgelet part detectors. In: *ICCV '05: Proceedings of the 10th IEEE Int. Conf. on Computer Vision (ICCV'05)* (Vol. 1, pp. 90–97). Washington, DC: IEEE Computer Society.

25. Kölsch, M., & Turk, M., (2004). Robust hand detection. In: *Proceedings of the Sixth International Conference on Automatic Face and Gesture Recognition* (pp. 614–619). Seoul: IEEE.

26. Dollar, P., Wojek, C., Schiele, B., & Perona, P., (2012). Pedestrian detection: An evaluation of the state of the art. *IEEE Trans. Pattern Anal. Mach. Intell., 34*, 743–761. doi: 10.1109/TPAMI.2011.155.

27. Bengio, Y., (2012). Deep learning of representations for unsupervised and transfer learning. In: Guyon, I., Dror, G., Lemaire, V., Taylor, G. W., & Silver, D. L., (eds.), *ICML Unsupervised and Transfer Learning: JMLR Proceedings* (Vol. 27, pp. 17–36). Bellevue: JMLR.Org.

28. Mesnil, G., Dauphin, Y., Glorot, X., Rifai, S., Bengio, Y., Goodfellow, I. J., et al., (2012). Unsupervised and transfer learning challenge: A deep learning approach. In: Guyon, I., Dror, G., Lemaire, V., Taylor, G., & Silver, D., (eds.), *JMLR W& CP: Proceedings of the Unsupervised and Transfer Learning Challenge and Workshop* (Vol. 27, pp. 97–110). Bellevue: JMLR.org.

29. Kotzias, D., Denil, M., Blunsom, P., & De Freitas, N., (2014). *Deep Multi-Instance Transfer Learning*. CoRR, abs/1411.3128.

30. Paletta, L., & Pinz, A., (2000). Active object recognition by view integration and reinforcement learning. *Rob. Auton. Syst., 31*, 71–86. doi: 10.1016/ S0921-8890(99)00079-2.

31. Correa, M., Hermosilla, G., Verschae, R., & Ruiz-Del-Solar, J., (2012). Human detection and identification by robots using thermal and visual information in domestic environments. *J. Intell. Robot Syst., 66*, 223–243. doi: 10.1007/s10846-011-9612-2.

32. Peng, B., Zhang, L., & Zhang, D., (2013). A survey of graph theoretical approaches to image segmentation. *Pattern Recognit., 46*, 1020–1038. doi: 10.1016/j. patcog.2012.09.015.

33. Aloimonos, J., Weiss, I., & Bandyopadhyay, A., (1988). Active vision. *Int. J. Comput. Vis., 1*, 333–356. doi: 10.1007/BF00133571.

34. Hoory, S., Shapira, T., Shabtai, A., & Elovici, Y., (2020). *Dynamic Adversarial Patch for Evading Object Detection Models.* arXiv preprint arXiv:2010.13070v1.

35. Thys, S., Ranst, W. V., & Goedemé, T., (2019). *Fooling Automated Surveillance Cameras: Adversarial Patches to Attack Person Detection.* arXiv preprint arXiv:1904.08653v1.

CHAPTER 4

Design of a Flow Detection Sensor for Single Phase Flow Through Porous Media

TALHA KHAN[1] and ANUSTUP CHATTERJEE[2]

[1]Department of Mechanical Engineering, Wichita State University, Wichita, United States of America

[2]Department of Mechanical Engineering, Techno International Newtown, Kolkata, West Bengal, India

ABSTRACT

A porous media consists of a 3D intricate network of pores that are arranged either in a symmetric or non-symmetric pattern, allowing for the fluid to pass through the void spaces. The larger the pores size or the porosity of the medium, the higher the permeability of the fluid through the porous medium [8]. Porosity, in this case, is defined as the ratio of the volume of the void space to the total volume of the matrix, and permeability is defined as the measure of the porous media to allow the fluid to flow through the pores. As such, the transportation of fluid through a porous medium is of high interest in understanding the effects of material porosity and permeability in heat, mass, and moisture transport in real-life situations such as percolation of water to the underground water table, contaminant transport in aquifers, chromatography separations, etc., where the natural porous media govern a plethora of interacting and dynamic processes and as such a better understanding of pore scale transport phenomena has become necessary [3]. The numerical modeling of fluid transport

System Design Using the Internet of Things with Deep Learning Applications.
Arpan Deyasi, Angsuman Sarkar, and Soumen Santra (Eds)

through a natural porous media using commercial software, COMSOL Multiphysics, to emulate these natural processes is used for assisting engineers and researchers in comprehending the ill effects of these natural processes, if any, and at the same time replicating these natural processes artificially in places where similar end effects are required. With a proper understanding of the fluid flow in these porous media, efficient systems can be designed keeping in mind the recent energy crisis and the need for sustainable development.

4.1 INTRODUCTION

A porous medium, by definition, is any material that is composed of a dense matrix and a network of interconnected voids. The matrix of the porous material comprises the skeletal portion of the material. The pores of a porous medium are typically occupied by a fluid (liquid or gas). One or more fluids will flow through the material due to the interconnectedness of the voids or the pores [11]. The fluid can flow through the pores can be in a single phase or in a multiphase [9]. In sponges, the solid matrix, and the porous network (called pore space) are continuous in sponges resulting in the formation of interpenetrating phenomena. Fluid flow in a porous media is caused by the differential viscosity, gravitational forces, and capillary action through the pore spaces [10].

The distribution of pores in a normal porous medium is irregular in terms of form and scale. Sand, sandstone, shale, rye bread, wood, and the human lung are examples of porous media in nature [12]. Subsequently, the determination of flow through a porous media is of extreme importance for characterization purposes. In these cases, a flow sensor comes into play when determining the flow rate of any single-phase flow in a porous medium. A flow sensor is an electronic device that is used to measure or regulate the rate of flow of any liquid or gas in a flow channel. Flow sensors are usually used in conjunction with gauges to calibrate the measurements and for proper recording for future purposes. Flow sensors can be of two major types depending upon the use, viz., differential pressure flow sensor and mass flow sensor. However, the most common parameters controlling the type of flow sensor used are material type, pressure, and required accuracy. Differential flow sensors like venturi tubes, rotameters, and orifice plates are predominantly used due to their performance accuracy and versatility of applications, whereas the mass flow sensors are used in automotive industries where they are used for measuring the air intake systems of an internal combustion engine.

The divergence between the solver used for the numerical simulation of a flow through porous media and the real flow structure determines the way in which the flow through the porous media is modeled [1]. When the distance between the object and the observer is small, the flow through the medium is perceived to either consist of single or multiple channels or single or multiple closed cavities. For such types of flow, the conventional rules for flow dynamics and convective heat transfer is applicable; as such, the utilization of a flow sensor is necessary for determining the flow rate in a porous media to eliminate the utilization of expensive experimental setups for characterization purposes.

The complexities of the path of the flow in a porous media rule out the traditional solutions where the conventional techniques of fluid flow and heat transfer are predominant. Volume aggregation and global measurements (e.g., permeability, conductivity) are helpful in defining and simplifying the flow in this maximum. Examples of filtration, geomechanics, soil mechanics, rock mechanics, petroleum engineering, geosciences (hydrogeology, petroleum geology, geophysics), biology, biophysics, material science, and others all use the principle of porous media [5]. Fluid flow through porous media is a topic of widespread interest that has spawned its own area of research. In synonyms with the development of research in fluid flow in porous media, there is the simultaneous development of widespread application of sensors in porous media flow.

The characterization of a porous medium is usually done on the basis of two main methods, viz., porosity and permeability, where the porosity of a porous structure is defined as the fraction of the total volume of the medium that is occupied by void space, where permeability is defined as the ability of a porous matrix to allow flow through it. These two major factors are taken into consideration when determining the design of the flow sensor for fluid flow in a porous media [6].

4.1.1 DARCY'S LAW

This section discusses the modification of the momentum equation for a porous medium which is the analog of the Navier-Stokes equation [1] and [13]. In 1855, a French hydraulic engineer named Henri Darcy worked on a series of experiments to determine the flow rate of water through layers of sand and its relation to the loss of pressure head along the path of flow. However, Darcy's law is applicable to low flows with low velocity and to flows without boundary shear flows [2].

For the studies, Darcy used a vertical column with a water inlet from one end and an outlet from the other. The setup had reservoirs to replenish and maintain the constant water levels and also the pressures at the inlet and outlet (denoted h_1 and h_2) (Figure 4.1). The experiments were performed using different sand packings and as well as sets of tests with the same packing but using different column heights.

Darcy determined a tube's discharge by calculating the amount of water that flows out the outlet over a set period of time (e.g., 1 minute). His studies were designed to look at the fundamental factors that influence groundwater flow. It employs a "Darcy channel," a sediment-filled tube from which water will flow. Water flows from a raised tank into the Darcy tube's inlet, into the sediment, and out the tube's outlet. Darcy's results quantified the relationships between volumetric groundwater flow rate, moving forces, and aquifer properties, laying the groundwork for modern hydrogeology. Darcy's experiments revealed proportionalities between the flow of water, Q, through the laboratory "aquifer" and the properties of several experimental devices [14].

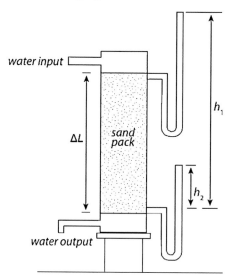

FIGURE 4.1   Schematic of Darcy's experimental setup.

Source: Maureen Feineman, Pennsylvania State University.

i.   Q(Flow rate) is directly proportional to the difference in water levels between the inlet and outlet, $h_1 - h_2 = \Delta h$: $Q \propto \Delta h$.

ii. Q(Flow rate) is proportional to the cross-sectional area of the vertical column: $Q \propto A$.

iii. Q(Flow rate) is inversely proportional to the length of the column: $Q \propto 1/\Delta L$.

Combining the above proportionalities leads to Darcy's Law, which is given as:

$$Q = KA(\Delta h/\Delta L)$$

where; A is the cross-sectional area; L is the length of the column; K is the proportionality constant which is denoted as hydraulic conductivity; $(h_1 - h_2)/l$ is the hydraulic gradient.

The permeability of any porous medium is determined by its porosity, grain size, and interconnection of the pores. As such, Darcy's experiment is dependent upon the type of grain bed used.

4.1.2   FLOW SENSORS/METERS

Flows can be open channel type and closed channel type. When a flowing stream has a free, unrestrained surface, it is called an open channel flow, whereas a flow in a channel is one occurring in a closed system or a conduit and the flow is a pressure-driven one. Flow in the water supply of pipes is the most common example of a closed channel flow, while on the other hand, an open channel flow is one that is usually evident in sand beds, usually gravity-driven flow.

4.1.2.1   DIFFERENTIAL PRESSURE FLOW SENSOR

The flow is measured in a differential pressure drop set up by calculating the drop in head caused by a barrier in the fluid flow path. This drop is evaluated using a signal which is a function of the square of the speed of the flow calculated using Bernoulli's equation. The most widely used differential flow meters are discussed in subsections.

4.1.2.1.1   Orifice Plates

The difference in pressure between the upstream and downstream sides of a partially obstructed channel is used to calculate fluid flow with an orifice plate. The flow-obstructing plate forces the flowing fluid to contract by creating a carefully calculated barrier in the pipe. Orifice plates are simple, inexpensive, and may be made in almost any material for virtually any use. Figure 4.2 depicts the schematic of an orifice plate setup.

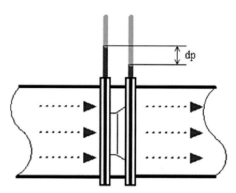

FIGURE 4.2   Orifice plate.
Source: Reprinted from Ref. [16]. Open access.

4.1.2.1.2   *Venturi Tube*

The venturi-type flowmeter is applicable where low-pressure drops are needed as compared to orifice plates, due to the simplicity in design. The rate of flow is determined by a reduction in the cross-sectional flow area, thereby causing a pressure difference, as shown in Figure 4.3. After passing through the neck region of the tube, the fluid passes through a pressure recovery section where 80% of the head is developed.

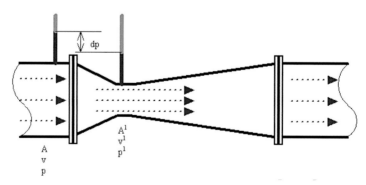

FIGURE 4.3   Venturi tube.
Source: Reprinted from Ref. [16]. Open access.

4.1.2.1.3   Flow Nozzle

In industrial applications, flow nozzles are often used as measurement elements for air and gas flow, as shown in Figure 4.4.

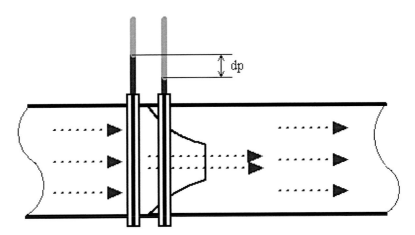

FIGURE 4.4   Flow nozzle.

Source: Reprinted from Ref. [16]. Open access.

4.1.2.2   MASS FLOW SENSOR

Mass meters directly measure the mass flow rate in a flow channel. The different types of mass flow meters are given in subsections.

4.1.2.2.1   Thermal Flow Meter

The thermal mass flowmeter does not rely on density, strain, or viscosity to work. Thermal meters use a heated sensing system that is separated from the fluid flow direction and emits heat from the sensing element to the flow stream. The conducted heat is exactly proportional to the mass flow rate, and the temperature differential is computed to mass flow.

The accuracy of the thermal mass flow system is determined by the process's calibrations as well as differences in the fluid's temperature, pressure, flow rate, heat power, and viscosity.

4.1.2.2.2 Coriolis Flowmeter

The Coriolis Mass Flowmeter measures the volume of mass flowing through the element using the Coriolis effect. The fluid to be analyzed is forced to vibrate in an angular harmonic oscillation through a U-shaped conduit. The tubes can deform due to Coriolis forces, adding an extra vibration dimension to the oscillation. This extra part triggers a phase change in certain areas of the tubes, which can be detected using sensors.

4.2 NUMERICAL MODELING USING COMSOL

Two different porosity configuration cases are being numerically modeled and analyzed so that a comparison can be made between the effect of pore size on porosity and permeability, viz., aligned and staggered grid, as shown in Figure 4.5 for both arrangements. This is being done to ascertain the effect of the pore size and its arrangement on the flow dynamics of the porous media. As such, this can be used for the design and manufacturing of metal foams and other additively manufactured components.

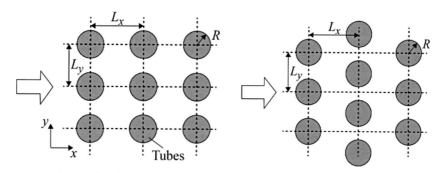

FIGURE 4.5 Aligned tube bank (left); and staggered tube bank arrangement (right).

4.2.1 GEOMETRY OF ALIGNED TUBE ARRANGEMENT

Modeling of the aligned tube arrangement was made using COMSOL Multiphysics Geometry Modeler using two different dimensional configurations for the aligned tube arrangements, mentioned below. Figures 4.6 and 4.7 show the geometry created for aligned tube arrangement for both dimensional configurations.

$$L_x = L_y = 4 \text{ mm}$$

$$R = 0.5 \text{ mm}$$

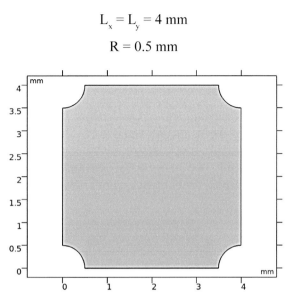

FIGURE 4.6   Geometry of aligned test case with Lx = Ly = 4 mm.

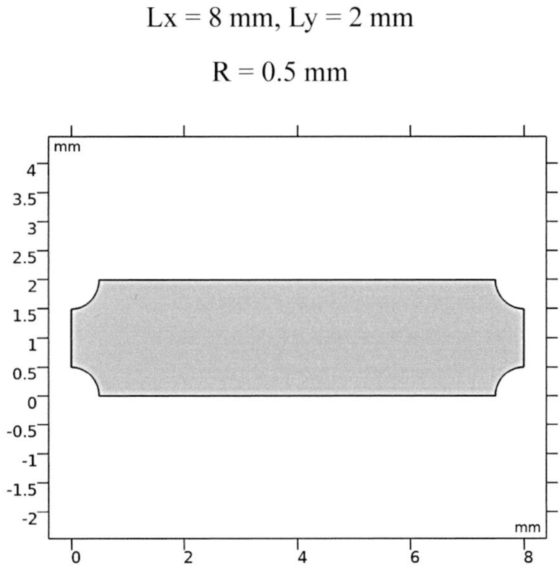

$$Lx = 8 \text{ mm}, Ly = 2 \text{ mm}$$

$$R = 0.5 \text{ mm}$$

FIGURE 4.7   Geometry of aligned test case with Lx = 8 mm, Ly = 2 mm.

The fluid property of the engine oil considered is as follows for setting up the test case:

$$\mu = 0.5 \text{ Pa-s}$$

$$\text{Density } \rho = 850 \text{ kg/m}^3$$

4.2.2   MESHING OF THE GEOMETRY AND CASE SETUP

The mesh of the geometry was created using COMSOL Multiphysics Default Mesher with an extra fine mesh size. This was done to capture the fluid flow properties at the edges of the tube where due to a change of geometry, there is eddy and micro-eddy generation. Figures 4.8 and 4.9 show the mesh generated for both configurations for aligned tube banks.

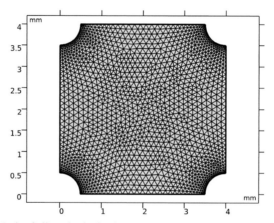

FIGURE 4.8   Mesh of aligned tube bank arrangement with Lx = Ly = 4 mm.

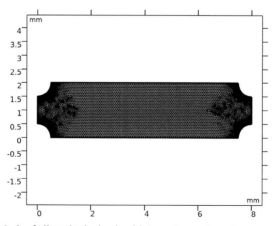

FIGURE 4.9   Mesh of aligned tube bank with Lx = 8 mm, Ly = 2 mm.

For setting up the case, a laminar single-phase creeping flow was taken into consideration. Inlet pressure was taken to be 0.1 Pa at the inlet, and the pressure at the outlet was taken to be 0 Pa to generate a pressure gradient-driven flow. A periodic flow condition was taken to be on the top, and bottom edges of the geometry, and the curved edges depicting the tubes in 2D were taken to be as a wall, and the path of the flow is considered to be in the x-direction with the inlet being at edge located at the origin. For simulating the case, a steady-state solver was employed.

4.2.2.1   GEOMETRY OF STAGGERED TUBE ARRANGEMENT

Modeling of the aligned tube arrangement was made using COMSOL Geometry Modeler using two different dimensions for the staggered tube arrangement, mentioned below. Figures 4.10 and 4.11 depict the model created for symmetric staggered tube arrangement.

$$Lx = Ly = 4 \text{ mm}$$
$$R = 0.5 \text{ mm}$$

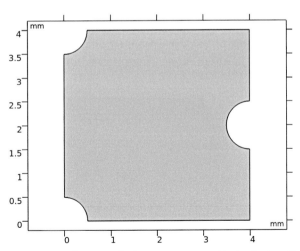

FIGURE 4.10   Geometry of staggered tube arrangement with the first configuration.

$$Lx = 8 \text{ mm}, Ly = 2 \text{ mm}$$
$$R = 0.5 \text{ mm}$$

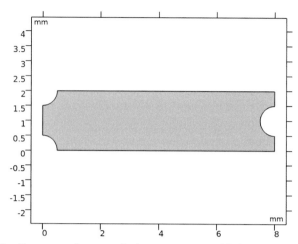

FIGURE 4.11   Geometry of staggered tube arrangement with the second configuration.

4.2.2.2 MESH OF STAGGERED TUBE ARRANGEMENT AND CASE SETUP

As in the case of aligned tube arrangement, the mesh of the staggered tube geometry was created using COMSOL Multiphysics Default Mesher with extra fine mesh size. This has been due to capturing the micro eddy generation at the places of change of path to the flow of the fluid, which enhances the mixing of the flow. Figures 4.12 and 4.13 show the mesh generated for both configurations of staggered tube bank arrangement.

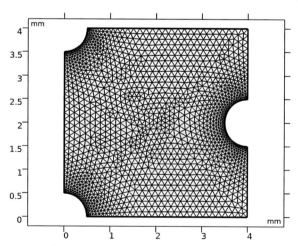

FIGURE 4.12   Mesh of staggered tube bank arrangement with first configuration.

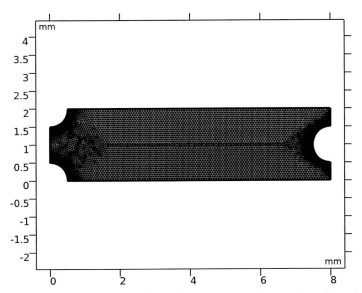

FIGURE 4.13   Mesh of staggered tube bank arrangement with second configuration.

A laminar single-phase creeping flow was considered when setting up the case. To produce a pressure gradient-induced flow, the inlet pressure was set to 0.1 Pa at the inlet, and the outlet pressure was set to 0 Pa at the outlet. As in the case of the aligned tube bank flow, the x-direction of flow is considered, with the inlet edge being the leftmost edge of the geometry. On the top and bottom sides of the geometry, a periodic flow condition was assumed, and the bent edges representing the channels in 2D were assumed to be walls. A steady-state solver was used to simulate the scenario.

4.2.3   RATE OF RISE FLOW METER FOR POROUS MEDIA FLOW MEASUREMENT

The rate of rise flow meter is a manual flow sensing device that is used to evaluate the rate of rise of any flow in a porous media. The rate-of-rise (RoR) method determines flow by measuring the time rate of change of height of a fluid flow in a porous media. This method is more of a visual method of detecting the flow of a fluid through a porous media where it involves the specimen being dipped in the sample solution, and then calculating the rate of rise of the height of the solution in the sample visually with respect to time. This method is mostly applied in wick structures to determine the rate of wicking, where wicking is defined as the rate of movement of the fluid through the capillary action from the inside of a porous wick structure. Figure 4.14 shows a typical setup used for the rate of rise experiment to determine the permeability of a sample.

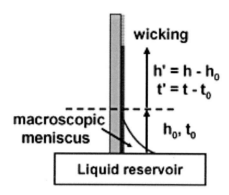

FIGURE 4.14   Rate of rise flow measurement setup.

Source: Nam et al. [15].

4.3   RESULTS AND DISCUSSION

4.3.1   *RESULTS OF ALIGNED TUBE BANK ARRANGEMENT*

Due to the complexity of the problem, the problem cannot be solved accurately by analytical means and be used for comparison. As such, the velocity and pressure contours are analyzed. Also, the velocity has been plotted to determine the maximum velocity during the flow. Figures 4.15–4.17 depict the profiles for aligned tube bank arrangement with Lx = Ly = 4 mm, whereas, Figures 4.18–4.20 show the profiles for Lx = 8 mm, Ly = 2 mm for the same tube arrangement.

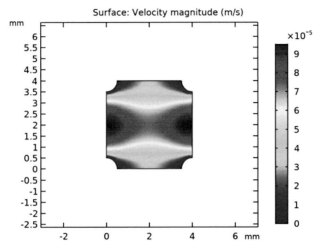

FIGURE 4.15   Velocity contours with Lx = Ly = 4 mm.

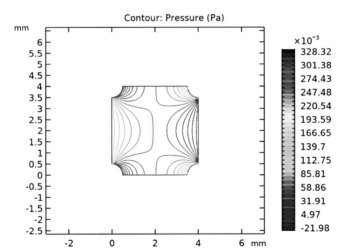

FIGURE 4.16   Pressure contours with Lx = Ly = 4 mm.

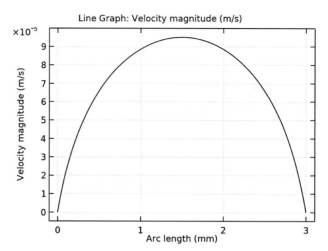

FIGURE 4.17   Velocity magnitude.

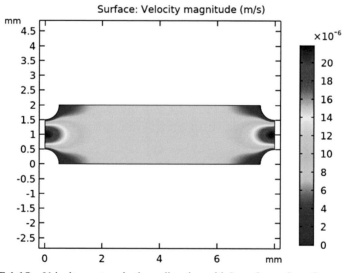

FIGURE 4.18   Velocity contour in the x-direction with Lx = 8 mm, Ly = 2 mm.

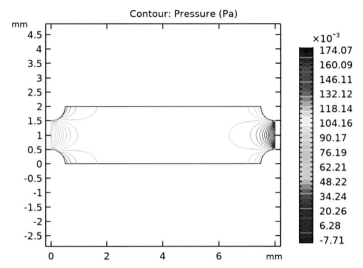

FIGURE 4.19   Pressure contour for x-direction of flow with Lx = 8 mm, Ly = 2 mm.

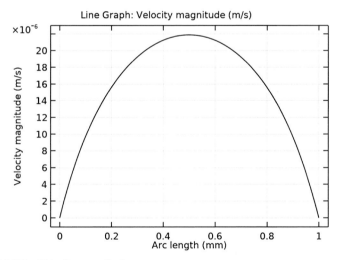

FIGURE 4.20   Velocity magnitude.

4.3.2   *RESULTS OF STAGGERED TUBE BANK ARRANGEMENT*

Similarly, as in the case of aligned tube bank arrangement, the velocity and pressure contours and velocity profiles have been plotted for the flow.

Figures 4.21–4.23 are for staggered tube arrangement for Lx = Ly = 4 mm whereas, Figures 4.24–4.26 are for staggered tube arrangement with the configuration of Lx = 8 mm, Ly = 2 mm.

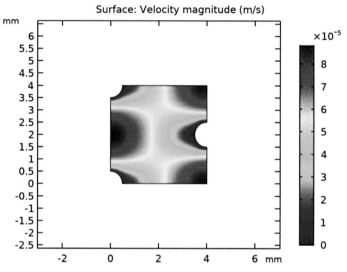

FIGURE 4.21   Velocity for staggered tube arrangement with Lx = Ly = 4 mm.

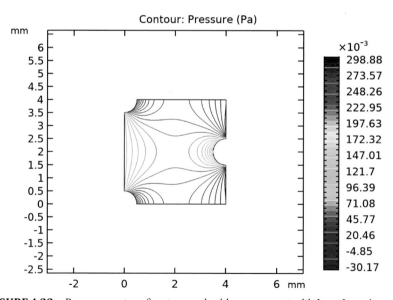

FIGURE 4.22   Pressure contour for staggered grid arrangement with Lx = Ly = 4 mm.

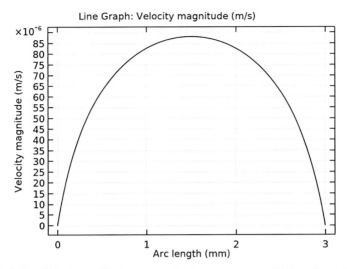

FIGURE 4.23   Velocity profile for staggered tube arrangement with Lx = Ly = 4 mm.

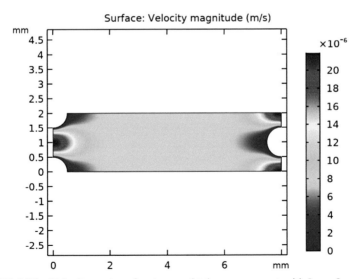

FIGURE 4.24   Velocity contour for staggered tube arrangement with Lx = 8 mm, Ly = 2 mm.

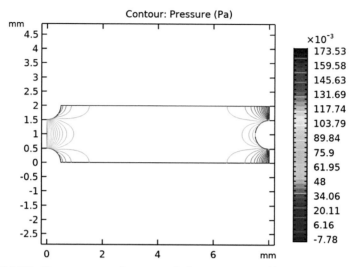

FIGURE 4.25   Pressure contour for staggered tube arrangement with Lx = 8 mm, Ly = 2 mm.

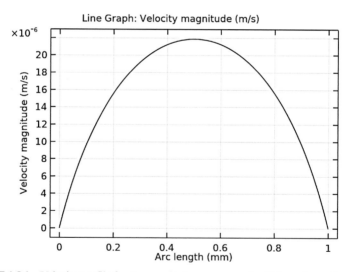

FIGURE 4.26   Velocity profile for staggered tube arrangement with Lx = 8 mm, Ly = 2 mm.

4.4   CONCLUSION

The permeability of the aligned tube bank is greater than the staggered tube path because the path of the flow of the fluid in the aligned tube bank is less tortuous, i.e., the fluid flow path in the aligned grid has less number of curves and flow path changes as compared to the staggered grid, where the fluid flow path has bends to the path of the flow. The staggered grid has flow separation at multiple points at the leading edge of the staggered arrangement of the cylinders; as such, the permeability reduces. Also, for Lx = Ly = 4 mm dimension of the aligned tube bank, as compared to Lx = 8 mm and Ly = 2 mm, from the velocity profiles, it is seen that for the symmetric dimension, the velocity is more tortuous as compared to the non-symmetric dimension due to the fact that the flow develops along the length in the x direction and as such the flow settles due to no change in flow path subsequently making the flow magnitude constant. Similarly, for the staggered tube arrangement for the symmetric arrangement, the change velocity magnitude is higher at the central zone as compared to the non-symmetric arrangement due to flow development and settling.

The rate of rise of fluid or the capillary action in porous media is used in the thermal management system of high-power density semiconductor devices, such as microprocessors, power switching devices, and lasers. As such, these porous media, or wick structures, are being used as passive cooling devices as they have the ability to handle very high heat fluxes and offer effective thermal conductivities better than conventional cooling devices. However, the porous media design has to be characterized as per the heat removal requirement. Therefore the rate of the rising flow meter has been aptly characterized as per the requirement of the efficiency of heat removal of the porous media.

KEYWORDS

- **conventional cooling devices**
- **Darcy's law**
- **micro eddy generation**
- **permeability**
- **porosity**
- **porous medium**
- **semiconductor**
- **tortuous**

REFERENCES

1. Nield, D. A., & Bejan, A., (2006). *Convection in Porous Media* (Vol. 3). New York: Springer.
2. Ho, C. K., & Webb, S. W., (2006). *Gas Transport in Porous Media* (Vol. 20). Springer Science & Business Media.
3. Dullien, F. A., (2012). *Porous Media: Fluid Transport and Pore Structure*. Academic Press.
4. De Boer, R., (2012). *Theory of Porous Media: Highlights in Historical Development and Current State*. Springer Science & Business Media.
5. Scheidegger, A. E., (2020). *The Physics of Flow Through Porous Media*. University of Toronto Press.
6. Adler, P. M., Jacquin, C. G., & Quiblier, J. A., (1990). Flow in simulated porous media. *International Journal of Multiphase Flow, 16*(4), 691–712.
7. Vafai, K., (2015). *Handbook of Porous Media*. CRC Press.
8. Ingham, D. B., & Pop, I., (1998). *Transport Phenomena in Porous Media*. Elsevier.
9. Bear, J., (2013). *Dynamics of Fluids in Porous Media*. Courier Corporation.
10. Satter, A., & Iqbal, G. M., (2016). Fundamentals of fluid flow through porous media. *Reservoir Engineering,* 155–169.
11. Collins, R. E., (1976). *Flow of Fluids Through Porous Materials*. United States. https://www.osti.gov/biblio/7099752.
12. Philip, J. R., (1970). Flow in porous media. *Annual Review of Fluid Mechanics, 2*(1), 177–204.
13. Durst, F., Haas, R., & Interthal, W., (1987). The nature of flows through porous media. *Journal of Non-Newtonian Fluid Mechanics, 22*(2), 169–189.
14. https://ocw.snu.ac.kr/sites/default/files/MATERIAL/7756.pdf (accessed on 11 October 2022).
15. Nam, Y., Sharratt, S., Byon, C., Kim, S. J., & Ju, Y. S., (2010). Fabrication and characterization of the capillary performance of superhydrophilic Cu micropost arrays. *Journal of Microelectromechanical Systems*, *19*(3), pp. 581–588.
16. Engineering ToolBox, (2003). *Fluid Flowmeters - Comparing Types*. [online] Available at: https://www.engineeringtoolbox.com/flow-meters-d_493.html [Accessed 5 April 2023].

CHAPTER 5

Deep Neural Network-Based Brain Tumor Segmentation

SAYAN CHATTERJEE,[1] MOITRI CHAKRABORTY,[2] AKASH MAITY,[2] and MAINAK BISWAS[2]

[1]Senior System Engineer, Cognizant, Technology Solutions, Kolkata, West Bengal, India

[2]Department of Electrical Engineering, Techno International New Town, Kolkata, West Bengal, India

ABSTRACT

Tumor is a dangerous disease that kills thousands of people every day in India. The key cause is abrupt organ mass development. Tumors are divided into two types: benign and malignant. The main issue is that everyone's size and shape are different. In order to discover this at an early stage, perfect segmentation is critical. Since manual segmentation varies between observers, it is sometimes inconclusive. This work introduces a novel deep neural network-based methodology for segmenting brain tumors and their sub-regions from multimodal MRI images, as well as survival prediction using characteristics generated from segmented tumor sub-regions and clinical characteristics. This study defines a computer-aided diagnostic approach. This section defines a computer-aided diagnostic approach that uses a deep learning-based technique called TransResV-Net, which is a V-net extension, to segment tumor cells. The revised V-Net outperformed the current finest practice strategies. V-net is more successful than the traditional network for pixel-based segmentation, and it is one of the

System Design Using the Internet of Things with Deep Learning Applications.
Arpan Deyasi, Angsuman Sarkar, and Soumen Santra (Eds)
© 2024 Apple Academic Press, Inc. Co-published with CRC Press (Taylor & Francis)

simplest methods for segmenting tiny data sets. This methodology is used to partition usable data collecting. This model is applied to the usable BRATS 2015, Kaggle data set comprising 120 patients' MRI images to apply the segmentation. Apart from this, various databases are available. But this BRATS 2015 is chosen for this experiment. This experiment is being conducted for 20 epochs, and the result is being compared with the supplied ground truth, and the segmented outcome is assessed using metrics such as intersection-over-union; the dice coefficient was reported as 0.72 and 0.87, respectively. As compared to well-known and benchmark approaches, the result demonstrated a significant difference. Researchers in the fields of bioinformatics and medicine will benefit greatly from the high-performing TransResV-Net.

5.1 INTRODUCTION

One of the most sensitive and varied organs in the human body is the brain, composed of several cells that function in tandem. A brain tumor develops as these cells congregate in or near the brain due to unregulated cell division [1]. These unwanted cells have the potential to kill other healthy cells while also interfering with the brain's natural functionality.

Brain tumors are generally classified into:

i. Low-Grade (Grade I or II) or Benign Tumors: These cancers are not very dangerous and cannot spread to other areas of the body, making them non-cancerous. They are developed in the brain and develop at a slow pace.

ii. High-Grade (Grade III or IV) or Malignant Tumors: These cancers are more aggressive than benign tumors and develop quickly, making them cancerous. They can be central, which means they originate in the brain, or secondary, which means they originate elsewhere in the body and subsequently spread to the brain [2].

Brain tumor segmentation is a critical step in medical image processing. Early detection of brain tumors improves therapeutic options and raises patients' chances of recovery. The manual segmentation of brain tumors from vast volumes of MRI data generated in clinical practice is a hard and time-consuming technique [3]. The most lethal sickness is a tumor, and if it is in the brain, it is considerably more lethal. A biopsy can be performed for brain tumors in difficult-to-reach locations or in highly vulnerable parts of the brain that might be harmed by a more invasive technique; your neurosurgeon will drill a small hole into your brain. The cavity is then pierced with a small needle. The needle is used to scrap tissue and is often guided by CT or MRI scanning [1]. The mortality rate in India is almost 1,300 people per day [4]. Screening is very necessary to survive this. The main stage is to make a CAD system. For a long time, healthcare clinicians and researchers have focused on MRI (magnetic resonance imaging) for the diagnosis and further modeling of tumor progression during the process of treatment of brain tumors. This approach is favored mostly because of the images, which are of good resolution, which aid in collecting a large volume of knowledge about the subject. Support vector machines (SVM) and Neural Networks are two other methods that have gained a lot of traction in this area over the last decade. Figure 5.1 shows the various stages of a brain tumor [5], where WHO graded these three images as Grade II, III, and IV (from (a), (b), and (c)) [6].

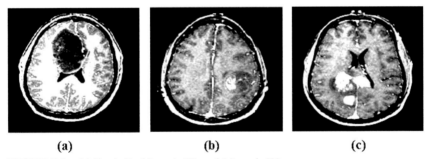

(a)   **(b)**   **(c)**

FIGURE 5.1   (a) Grade II; (b) grade III; and (c) grade IV.

For regression of brain tumors in MRIs, the following steps are [7]:

1. **Data Acquisition of Brain MRIs:** According to the World Health Organization classification system for identifying brain tumors, there are more than 120 different forms of brain tumors, most of which differ in origin.

2. **Image Segmentation:** It is the method of segregating the various types of MRIs. Normal brain components such as gray matter, white matter, and cerebrospinal fluid, as well as the skull, were separated from the actual tumor. This is done to make the tumor

study easier in the approaching phases. This is accomplished by the use of the Fuzzy-C means clustering algorithm, which divides the picture into five sections or clusters.

3. **Feature Extraction and Reduction:** DWT (discrete wavelet transform) principal component analysis is used for minimization and is utilized for feature selection. DWT can extract the most relevant characteristics at different directions and scales in a hierarchical manner; they provide localized time-frequency information about a signal by employing cascaded filter banks comprising both high and low-pass filters. The PCAs are then used to reduce the initially extracted features to lower-dimensional feature vectors.

4. **Regression Using DNN:** This is done using the 7-fold cross-validation process for creating and training the neural network of a few hidden structures [8].

However, great progress has recently been made in the application of deep learning in the diagnosis and analysis of brain cancers. This is owing to the ease with which complicated relationships may be expressed without the requirement for as many nodes as SVMs and KNNs need. TransResV-net, a deep learning-based network, is used in this case to separate the tumor region from the MRI image on the data set [9]. The key benefit of this neural network is that it is a pixel-based segmentation approach, and it performs well with tiny data sets. Many studies on tumor segmentation have been conducted. Zhao et al. used a combination of Conditional Random Fields and a fully convolutional network to segment tumor cells [10]. Zhou et al. proposed 3D U-net to segment tumor cells using a combination of dilated convolution network and deep supervision [11]. Tewari et al. used deep transfer learning to classify the tumor, where accuracy is 100% [5]. Pradeep et al. segment the tumor cells using deep learning-based wavelet analysis [12]. Several classical methods are also adopted for the segmentation process [13], but the deep learning-based technique is better than the classical one.

Here deep neural network-based segmentation model is discussed based on a modification of V-net. The work is organized in such a way, at first, there is an "introduction," and then the "Methodology" is described. Then the "Data set" section is given. "Result analysis" is described, followed by "Experimental Setup," then "Conclusion" is there.

5.2   METHODOLOGY

Brain segmentation may be divided into three categories based on the amount of user engagement required:

1. **Methods for Manual Segmentation:** Manual segmentation is a time-consuming operation that is heavily reliant on the radiologist's expertise. This means that, in addition to the anatomical and physiological knowledge gained via preparation and experience, the radiologist must make use of the multi-modality information provided by MRIs. The radiologist examines numerous slices of the photos slice by slice, then diagnoses and sketches the tumor regions carefully and methodically. However, the outcomes can differ greatly from one another [14].

2. **Semi-Automatic Segmentation:** Human intervention is primarily needed for three purposes in the semi-automatic method: initialization, intervention, and assessment. During the initialization phase, an estimated region of possibility where the tumor may be situated is supplied for the automated algorithm to process, and pre-processing technique settings can be changed. By acquiring input and changing answers, the automated algorithms may then be adjusted to create the desired result. Furthermore, the customer will assess the findings and redo the whole procedure if they are not acceptable [15].

3. **Fully Automatic Segmentation:** Human intervention is not needed in fully automatic segmentation. In this approach, a mixture of AI and exact knowledge is being used to identify the separation issue. Automatic segregation is divided into two types: discriminative and generative. Discriminative approaches, in most cases, rely on supervised learning methods, which necessitate large data sets with a valid ground set. They try to figure out what the link is between the input picture and the ground reality. They are mostly determined by the features chosen and the extraction of features [16].

However, generative approaches produce likely probable maps by using previous information, such as the spatial scale of healthy tissues and the site of the tumor. Having said that, it is also worth noting that automated segmentation is difficult in the case of gliomas (a type of brain tumor). This is due to the uneven and unclear boundaries of the tumor, which pose a huge difficulty, especially against the traditional edge-based methods. Furthermore, the tumor's form, size, and placement in the tumor-bearing brain might vary from person to person [17].

Deep learning, also known as deep neural networks, is an AI-based function and a branch of machine learning that replicates the workings of the human brain in identifying expression, making decisions, and manipulating data, i.e., the capacity to learn from both unstructured and unlabeled data by constructing a feature hierarchy.

It has features such as image de-noising, pixel-based segmentation, regression, and registration that have piqued the interest of researchers all over the world. Deep Neural Network is a Deep Learning architecture that is frequently used for categorization all around the world. A neural network is a basic feed-forward route network in which data goes from the input layer to the output layer through numerous hidden layers. The structure of a conventional neural network is depicted in the diagram above.

The bulk of diagnostic evidence in clinical practice, such as MRI, is composed of three-dimensional pictures. While most image analysis approaches can only process 2D images, completely convolutional neural network (CNN) based image segmentation of 3D volumetric data is cutting-edge [18].

V-Net architecture is identical to U-Net architecture with a few exceptions. It is divided into the compression path, the expansion path, the expansion path (left part), and the horizontal connections. Let us discuss the following two parts a bit more elaborately:

- **Left:**
    - It is divided into phases that operate at different resolutions. Each step on the left is made up of one to three convolutional layers.
    - At each level, the input is processed by the convolutional layers, and the residual function is learned to be added to the last layer of that stage as an output.
    - A voxel is the representation of the value on a regular grid in 3D space. Volumetric kernels of 5*5*5 are used by the convolutions performed at each stage.
    - The number of feature channels is doubled at each level of the compression path, the resolution is reduced, and the size of the

resultant feature maps is half as a consequence, with a goal similar to that of pooling layers.
- Down-sampling assists in increasing the receptive field.

- **Right:**
  - To collect the required information, this section extracts the features and boosts the spatial support of the lower-resolution feature maps.
  - A de-convolutional operation is used at each level to enhance the size of the inputs, which are then followed by one to three convolutional layers.
  - The residual function, like the left part, is taught in the right part as well.
  - The final convolution layer produces two output feature maps, each of which is the same size as the input volume.
  - These two maps are the probable, possible segmentation of the front and back regions.

- **Horizontal Paths:**
  - Location information is lost in the left path, like U-net.
  - As a result, characteristics obtained from the left path's early phases are passed on to the right path via horizontal connections.
  - This helps in providing location information to the right part and consequently getting a better-quality of the final image.
  - And these connections do help in improving the convergence time of the model.

The primary V-Net model is split into two parts: encoder (feature extraction) and decoder (create the map to partition the job depending on the features) [19].

There is one similarity of this model with the conventional neural network, i.e., the encoder part is the same as a convolutional network. The input image passes through some steps, like two 3 x 3 convolution layers having the activation done by ReLU, and then from the feature map, the maximum pooling is done. Later on, the number of feature channels is increased by double its previous, and spatial measures are decreased by half [20].

In the case of the decoder part, the feature map calculation is the same as the encoder with two 3 x 3 convolution layers having the activation again the ReLU. The feature stream quantity and spatial are repeated in the decoder section to keep the picture the same size as the input.

In this way, the size is the same so that the result can easily be compared with the ground truth [9]. The structure is the extension of V net [21].

To concentrate on tumor cell segmentation using biomedical image processing, some modification is required on classical V-Net models, and the name is TransResV-Net. The proposed model is demonstrated here, with all possible definitions. Figure 5.2 is the schematic diagram of TransResV-Net.

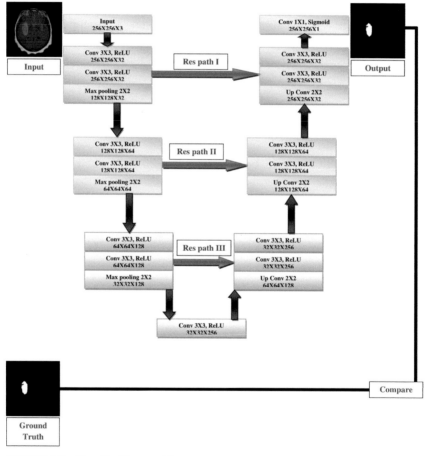

FIGURE 5.2   TransResV-net architecture.

5.3   DATASET

At the moment, precise interpretation of the effects of different images from brain tumor segmentation techniques is a daunting challenge.

However, the BRATS benchmark has become a commonly recognized benchmark for automated brain tumor segmentation. This enables remote comparison of various glioma segmentation approaches using standard datasets.

The suggested technique divides the input magnetic resonance image (MRI) of the tumor into many steps and finally feeds the 2nd pixel of all steps into the neural network. The DNN then marks these middle pixels and performs segmentation.

BRATS 2012 (2013, 2014, 2015, and ISLES 2015 and 2017 are among the eight large-scale benchmark datasets used.

The BRATS 2015 image data set contains MR images, which are taken from the cancer imaging archive (TCIA) [2]. The images are in .csv format.

And there are original images as well as masks of the corresponding images. The total number of patients is 120; among them, 10 patients' data need to be discarded because of the non-availability of proper cluster information. The sample images are given in Figure 5.3. There are four photos here, each having its own ground truth. Firth two images have a tumor, but the last two images do not have a tumor.

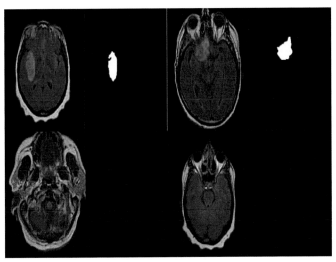

FIGURE 5.3   Data set: Upper images have the tumor cell with ground truth and lower images have no tumor cell with ground truth.

5.4   DICE COEFFICIENT

For three primary sections of the tumor, namely the entire tumor, core tumor, and active tumor, the findings are given via the online evaluation tool in the form of Dice Score, Sensitivity, and Specificity. The total tumor illustrates all of the tumor's components, the core tumor represents everything in the tumor, excluding edema, and the active tumor exhibits only the active cells [22].

The dice coefficient can be defined as twice the overlapped area by the total pixel number for both two images (image and ground truth). The equation is given as in Eqn. (5.1) and Figure 5.4.

$$D = 2d/b \tag{5.1}$$

where; D is the Dice coefficient; d is the overlapped area of both images; b is the total pixel number of both images.

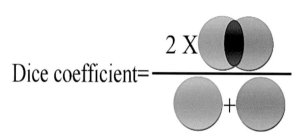

FIGURE 5.4   Dice coefficient.

5.5   INTERSECTION-OVER-UNION (IOU)

The overlapping region between the segmented output and the designated ground truth separated by the union of the two may be shown as intersection-over-union (IOU). In Eqn. (5.2) and Figure 5.5, the equation is as follows. The result ranges from 0 to 1, with 0 indicating no overlap and 1 indicating complete overlap [23–29].

$$X = z/v \tag{5.2}$$

where; X is the IoU; z is the overlapped area of both images; v is the union of both images.

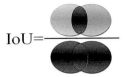

FIGURE 5.5 IoU.

The Dice coefficient is indeed very identical to the IoU factor. They are highly associated, which means that if one thinks model X is better than model Y at image segmentation, the other would agree.

They, such as the IoU, have a scale of 0 to 1, with 1 indicating the largest resemblance between expected and true.

Figure 5.6 shows the training loss and validation loss curves which should be decreased exponentially corresponding to the epoch while increasing.

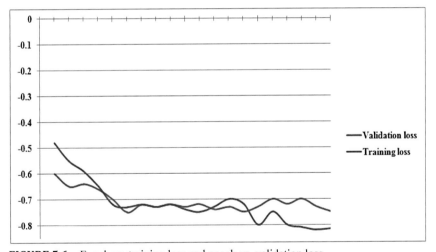

FIGURE 5.6 Epoch vs. training loss and epoch vs. validation loss.

It is very clear from the graph that the error decreases as the number of epochs increases. The errors should be minimized, and it happens. And Figure 5.7 shows the pictorial result of two input images and their corresponding ground truth. The segmented output and the ground truth are being compared, and the result is noted with respect to IoU and Dice efficiency. The value of IoU is 0.72. The result is given in Tables 5.1 and 5.2 with a comparison with the well-known methods.

TABLE 5.1   Comparison Table 1

Methods	Dice Coefficient
FCNN + CRF [25]	0.77
Fusing (FCNN+CRF) [25]	0.78
CRF + FCNN + post-process [25]	0.80
Fusing [10] + post-process [25]	0.82
DeepMedic + CRF [25]	0.85
FCNN(axial) + 3D CRF [25]	0.84
FCNN(coronal) + 3D CRF [25]	0.83
FCNN(sagittal) + 3D CRF [25]	0.82
FCNN + 3D CRF (fusing) [26]	0.84
Method [27]	0.86
Proposed	0.87

TABLE 5.2   Comparison Table 2

Methods	IoU
Method [29]	0.688
Proposed	0.72

From Figure 5.7, it is obvious that the technique is really high. It is also observed that when there is tumor exists, then only this method shows with appropriate size (Figure 5.7, first row), whereas in the second row, there is no such tumor in that image, and the method also shows that.

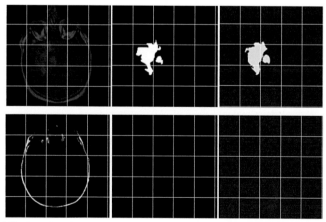

FIGURE 5.7   Result (chronologically: sample image, ground truth, segmented output, and row-wise).

This makes the framework most valuable to segment almost all types of brain tumors. It is also clear from Tables 5.1 and 5.2 that this method shows the best possible result.

5.6 CONCLUSION

Deep learning is used in this case. TransResV-Net is used to separate tumor cells from MR images of the brain. A total of 120 patient cases were used; among them, 110 cases were taken because of the non-availability of proper cluster information. It is very clear from Figure 5.6 that the training loss and validation loss are decreasing with respect to epochs. The main advantage is that it is pixel-based segmentation, and for small data sets, this method is the best.

ACKNOWLEDGMENTS

The authors are highly obliged to the Department of Electrical Engineering, Techno International New Town (formerly Techno India College of India), Kolkata, India, for their constant support and moral help. Though the work is not supported by any Foundation, the laboratory of the institute helped to do the work smoothly. We thank our colleagues from Techno International New Town (formerly Techno India College of India), who provided insight and expertise that greatly assisted the research, although they may not agree with all of the interpretations of this chapter.

KEYWORDS

- **brain tumor**
- **convolutional neural network**
- **deep learning**
- **magnetic resonance image**
- **support vector machines**
- **TransResV-net**
- **V net**

REFERENCES

1. Lumen Learning: https://courses.lumenlearning.com/ (accessed on 11 October 2022).
2. WebMD: https://www.webmd.com/cancer/default.htm/ (accessed on 11 October 2022).
3. Mayo Foundation for Medical Education and Research (MFMER). https://www.mayoclinic.org/ (accessed on 11 October 2022).
4. Shaw, E. G., Scheithauer, B. W., O'Fallon, J. R., Tazelaar, H. D., & Davis, D. H., (1992). Oligodendrogliomas: The mayo clinic experience. *Journal of Neurosurgery, 76*(3), 428–434.
5. Talo, M., Baloglu, U. B., Yıldırım, Ö., & Acharya, U. R., (2019). Application of deep transfer learning for automated brain abnormality classification using MR images. *Cognitive Systems Research, 54*, 176–188.
6. Cha, S., (2006). Update on brain tumor imaging: From anatomy to physiology. *American Journal of Neuroradiology, 27*(3), 475–487.
7. Bi, W. L., Hosny, A., Schabath, M. B., Giger, M. L., Birkbak, N. J., Mehrtash, A., & Aerts, H. J., (2019). Artificial intelligence in cancer imaging: Clinical challenges and applications. *CA: A Cancer Journal for Clinicians, 69*(2), 127–157.
8. Hayat, M. A., (2012). *Tumors of the Central Nervous System.* Berlin, Germany: Springer.
9. BRATS, (2015). *Data Set: Kaggle Page.* https://www.kaggle.com/mateuszbuda/lgg-mri-segmentation (accessed on 11 October 2022).
10. Zhao, X., Wu, Y., Song, G., Li, Z., Zhang, Y., & Fan, Y., (2018). A deep learning model integrating FCNNs and CRFs for brain tumor segmentation. *Medical Image Analysis, 43*, 98–111.
11. Zhou, T., Ruan, S., Hu, H., & Canu, S., (2019). Deep learning model integrating dilated convolution and deep supervision for brain tumor segmentation in multi-parametric MRI. In: *International Workshop on Machine Learning in Medical Imaging* (pp. 574–582). Springer, Cham.
12. Mallick, P. K., Ryu, S. H., Satapathy, S. K., Mishra, S., Nguyen, G. N., & Tiwari, P., (2019). Brain MRI image classification for cancer detection using deep wavelet autoencoder-based deep neural network. *IEEE Access, 7*, 46278–46287.
13. Chatterjee, S., Biswas, M., Maji, D., Ghosh, B. K., & Mandal, R. K., (2020). Logarithm similarity measure based automatic esophageal cancer detection using discrete wavelet transform. In: *Recent Trends and Advances in Artificial Intelligence and Internet of Things* (pp. 427–453). Springer, Cham.
14. Despotović, I., Goossens, B., & Philips, W., (2015). MRI segmentation of the human brain: Challenges, methods, and applications. *Computational and Mathematical Methods in Medicine, 2015.*
15. Chahal, P. K., Pandey, S., & Goel, S., (2020). A survey on brain tumor detection techniques for MR images. *Multimedia Tools and Applications, 79*, 21771–21814.
16. Dolz, J., Leroy, H. A., Reyns, N., Massoptier, L., & Vermandel, M., (2015). A fast and fully automated approach to segment optic nerves on MRI and its application to radiosurgery. In: *2015 IEEE 12th International Symposium on Biomedical Imaging (ISBI)* (pp. 1102–1105). IEEE.

17. Bahadure, N. B., Ray, A. K., & Thethi, H. P., (2017). Image analysis for MRI based brain tumor detection and feature extraction using biologically inspired BWT and SVM. *International Journal of Biomedical Imaging.*

18. Zhou, T., Ruan, S., & Canu, S., (2019). A review: Deep learning for medical image segmentation using multi-modality fusion. *Array, 3,* 100004.

19. Ronneberger, O., Fischer, P., & Brox, T., (2015). U-net: Convolutional networks for biomedical image segmentation. In: *International Conference on Medical Image Computing and Computer-Assisted Intervention* (pp. 234–241). Springer, Cham.

20. Towards Data Science: https://towardsdatascience.com (accessed on 11 October 2022).

21. Işın, A., Direkoğlu, C., & Şah, M., (2016). Review of MRI-based brain tumor image segmentation using deep learning methods. *Procedia Computer Science, 102,* 317–324.

22. Shamir, R. R., Duchin, Y., Kim, J., Sapiro, G., & Harel, N., (2019). *Continuous Dice Coefficient: A Method for Evaluating Probabilistic Segmentations.* arXiv preprint arXiv:1906.11031.

23. Rezatofighi, H., Tsoi, N., Gwak, J., Sadeghian, A., Reid, I., & Savarese, S., (2019). Generalized intersection over union: A metric and a loss for bounding box regression. In: *Proceedings of the IEEE/CVF Conference on Computer Vision and Pattern Recognition* (pp. 658–666).

24. Reza, S., Amin, O. B., & Hashem, M. M. A., (2020). TransResUNet: Improving U-net architecture for robust lungs segmentation in chest x-rays. In: *2020 IEEE Region 10 Symposium (TENSYMP)* (pp. 1592–1595). IEEE.

25. Kamnitsas, K., Ledig, C., Newcombe, V. F., Simpson, J. P., Kane, A. D., Menon, D. K., & Glocker, B., (2017). Efficient multi-scale 3D CNN with fully connected CRF for accurate brain lesion segmentation. *Medical Image Analysis, 36,* 61–78.

26. Kofler, F., Berger, C., Waldmannstetter, D., Lipkova, J., Ezhov, I., Tetteh, G., & Menze, B. H., (2020). Brats toolkit: Translating brats brain tumor segmentation algorithms into clinical and scientific practice. *Frontiers in Neuroscience, 14.*

27. Milletari, F., Navab, N., & Ahmadi, S. A., (2016). V-net: Fully convolutional neural networks for volumetric medical image segmentation. In: *2016 Fourth International Conference on 3D Vision (3DV)* (pp. 565–571). IEEE.

28. Gupta, D., Jain, S., Shaikh, F., & Singh, G., (2017). *Transfer Learning &The Art of Using Pre-Trained Models in Deep Learning.* Analytics Vidhya.

29. Haghighi, F., Taher, M. R. H., Zhou, Z., Gotway, M. B., & Liang, J., (2020). Learning semantics-enriched representation via self-discovery, self-classification, and self-restoration. In: *International Conference on Medical Image Computing and Computer-Assisted Intervention* (pp. 137–147). Springer, Cham.

CHAPTER 6

Voice Command-Based Real-Time Automated Home Security System with Google Assistance Technology

SOUMEN SANTRA,[1] PARTHA MUKHERJEE,[2] ARPAN DEYASI,[3] and ANGSUMAN SARKAR[4]

[1]Department of Computer Application, Techno International New Town, Kolkata, West Bengal, India

[2]Infiflex Technologies Pvt. Ltd., Kolkata, West Bengal, India

[3]Department of Electronics and Communication Engineering, RCC Institute of Information Technology, Kolkata, West Bengal, India

[4]Department of Electronics and Communication Engineering, Kalyani Government Engineering College, Nadia, West Bengal, India

ABSTRACT

In this chapter, a smart mirror-based home automation system is designed where voice commands (Google Assistance) in association with sensors are utilized for controlled regulatory measures. In this system, one channel is dedicated for motion sensing switching of the equipment and another dedicated for the light intensity-based switching of the equipment connected to the channel. In addition to turn-on and turn-off the lights as per voice request, it can also search the web for voice query and read out the results about the weather when asked for. It also smartly helps to reduce the energy consumption by switching off the device when not

System Design Using the Internet of Things with Deep Learning Applications.
Arpan Deyasi, Angsuman Sarkar, and Soumen Santra (Eds)

needed. This system can be operated through Android and IOS-based mobiles, tablets, and even through computers or any purchased Google home devices and self-created microcontroller-based docks that are associated with the owner's Google account and have the permission to trigger byte information to the authenticated server.

6.1 INTRODUCTION

Voice-operated switching control systems are the smarter version of the remote-controlled switching system, and can safely be applied inside a house, or a smaller unit, i.e., a room. In the last few years, demand of the smarter home is progressed [1, 2], and precisely for elderly people [3], speech recognition technology embedded with operation of electrical appliances become one key area of research. For trustworthy speech recognition process, extensive uses of DSP (digital signal processor) processors are mentioned by various workers [1, 4], but the cost of the system also increases, which makes it unaffordable to many of the consumers. However, enhancement of comfort level is another closely interrelated non-measurable parameter, which demands augmentation of the system with mobile devices. This will help to make a composite system which is easy to operate with lower manufacturing cost, and moreover, the control terminal becomes portable [5].

Recently, power line system at infrared (IR) wavelength [6] is proposed to design a home automation system, whereas simple Short Message Service technology is proposed by another group [7] for the same purpose. Similar work was earlier carried out by Jawarkar et al. [8] using appropriate programmed chip to sustain AT command. Voice recognition in this system plays a major role as per a few published articles [9–10]. However, the cost function and system portability for common user becomes a concern, which needs to be optimized, as well as to help elderly people, where too much system complexity ultimately leads to failure for day-to-day use. Here comes the novelty of the present proposal.

This proposal deals with an inexpensive system, where it uses Google Assistant (provided by Google to all their OS, and if necessary, it can be downloaded to other smartphone makers too like IOS by Apple), the IFTTT server and application, The Blynk application, and Google Cloud Console. In the hardware part, The NodeMCU 8266 micro-controller acts as the major communicator as it receives and sends data to the designed application and server, along with a relay board of 4/8 channel, who's regulatory voltage is 5 volts and has the capability for handling 10 Amp current through it. Natural English language is used to give commands to Google Assistance and receive data on the Blynk app. In the next section, the total process flow diagram is briefly explained and subsequent section explains the obtained result followed by a conclusion.

6.2   BRIEF OVERVIEW OF HOME AUTOMATION SYSTEM

Home automation system refers as smart home or intelligent home. Here we connect all units of components or part of home appliances through the art of science and technology. It is a system which tuning all kinds of features of home such as in environment [11]. It is one kind of system where it receives various kinds of input parameters and after assessing them gives output in terms of signals. Basically, it gathers all units within the internal environment and isolated these from the external environment. In highly tropical country the external environment parameters rapidly changes which can affect our home's internal environment and it's affected all the parts of home's appliances. Each and every appliances follow such instructions and is programmable in such a way where the changes of environmental parameters from external to internal affects their performances and reduce their lifetime execution period. Most of the home appliances such as refrigerator, air-conditioner, inverter, electric fans, smart television, evaporative cooler, washing machine, micro-oven, home theater systems, etc. In some cases, beyond the boundaries, accident protection and road safety systems [12] are also required to facilitate modern day needs. Most of them consider as embedded systems where then appliances running through the subroutines programmable within it. Here these subroutines are working on the basis of internal environmental variables or parameters. The indoor or home internal environmental parameters are air quality such as humidity, carbon dioxide and oxygen balance, sound quality, lighting comfort, odor, microorganisms, vibrations, thermal comfort, ergonomics, electromagnetic radiation, etc. These parameters are not the same inside and outside of a home in a tropical country where topographical features are changing rapidly throughout season-wise. Home is such type of combination of open and closed system where the internal and external parameters change rapidly. Home automation system is such kind of system which regulate or tuning internal parameters. These internal parameters are isolated from the external one and create an equilibrium stage between these two. Because most of the home appliances deteriorate their performance due to imbalance situations of internal and external parameters of the home. Home automation system is a kind of system which tunes the internal parameters of the home. It can tune input parameters, process parameters as well as output parameters also.

Here all parameters measure through the exchange of information in terms of signals. Signal is one of the forms of electromagnetic radiation which transfers from one system to another system. We know that three types of scattering of radiation are there: photoelectric, crompton, and coherent. To analyze this radiation, we need a device which absorbs this kind of radiation and gives a measurement of the level of radiation. Normally radiation represents a phenomenon about the environmental change happen for the incidents which changes through the parameters. These kinds of changes happen through the internal and external parameters of the home or a system. Based on these changes, a system defined as a closed or open system. Normally within the earth there is no existence of open system because every system surrounded by environment. So, by nature home is known as one kind of close system where internal and external parameters are connected with each other. As millions of radiation n form of scattering happen between the internal and external parameters of the home. So, to measure it we need a device which detect or measure these kinds of phenomenon or change of parameters of environment. We can use various kinds of sensors in home automation system because the sensor is one of the most efficient single points of contact for these parameters. Sensor is one kind of device where it detects the changes of measurement of parameters. Basically, each sensor works for a particular parameter. In-home automation system we use different kinds of sensors where each sensor performing different activities. Here below we mention some of the basic sensors which are used for a home automation system. We use heartbeat sensor, transmissive type IR sensor, reflective type IR sensor, rain sensor, touch sensor, temperature sensor, color sensor, alcohol sensor, PIR (passive infrared) sensor, gas sensor, smoke sensor, humidity sensor, water flow sensor, IR receiver, LDR (light dependent resistors) light sensor, photo translator light sensor, thermistor temperature sensor, gyroscope motion sensor, ultrasonic sensor, soil moisture sensor, proximity sensor, etc.

Here we give a short view for the above sensors regarding their functionalities:

1. **Heartbeat Sensor:** It is a plug-&-play device to measure the heart beat rate and pulse.

2. **Transmissive Type IR (Infrared) Sensor:** It is one kind of optical transmit and receive device which acts as an interrupter used to locate movement of object.

3. **Reflective Type IR (Infrared) Sensor:** It is another kind of motion detector sensor which detects signals coming from animal body.

4. **Rain Sensor:** It is a device which changes its modes through the rainfall, which alerts to close the windows and doors.

5. **Touch Sensor:** It is used to detect and store the record of touch coming from a living body.

6. **Temperature Sensor LM35:** It is a calibrated circuitry device which measures temperature and giving an analog voltage output.

7. **Color Sensor:** It is work upon photoelectric effect where it emits a light towards the object and receives the reflective one. By which it can detect the color of an object.

8. **Alcohol Sensor:** It is a mosquito gas sensor and known as MQ3 which detects the presence of ethanol in the environment. It is a chemi-resistor which detects any kind of gas leakage due to reaction of metal oxide semiconductor with the gas molecules.

9. **PIR (Passive Infrared) Sensor:** It is used as a security alarm or automatic light emission purpose for the detection of objects. It always emits IR signals and can measure that radiated IR signals from the object surface.

10. **Gas Sensor:** It's working principle just similar to alcohol sensor but it can detect any gas and emits a sound due to the presence of gas in environment [13] through voltage breakdown method.

11. **Smoke Sensor:** It can detect the presence of smoke in the environment by photoelectric or ionization method. Basically, it works as a fire alarm control purpose.

12. **Humidity Sensor:** It is a calibrated device BME280 which is known as a hygrometer to measure the amount of moisture present in the environment.

13. **Water Flow Sensor:** It contains a valve and a rotor which rotates when water flow through the valve and the sensor measures the number of rotations which proportion to amount of water flow and its speed. It is used in drainage system or washing machine or home water reservoir system.

14. **Water Level Sensor:** It contains a series of variable resistors which measure the resistance which is calibrated with the level of water. Here we use copper resistors which indicate the signal to the LED power. We can use this in water reservoir system.

15. **IR Receiver:** It is classified as active and passive IR sensor. It is a photodiode which detects an IR signal emitted from a transmitter. It is used to detect and analyze gas, water, flame, moisture, etc.

16. **LDR (Light Dependent Resistors) Sensor:** It contains a series of variable resistors which measure the intensity of light. We can use it to monitor light of corridor, bedroom, staircases of home automation system.

17. **Photo Transistor Sensor:** It can detect the level of light coming from the object. It can measure the difference between emitter and collector. It is used in lighting control and door and window locking system, etc.

18. **Thermistor Temperature Sensor:** It is a device which fully depends upon resistance where its proportion to change of temperature. It detects the temperature changes and controls the fan and air conditioner.

19. **Gyroscope Motion Sensor:** It detects motion and orientation of object and also control it. It is used in automatic door or window locking system.

20. **Accelerometer Sensor:** It can measure the acceleration and speed of coming object towards it. It is used in alarm or control of door-window-fan-air conditioning system.

21. **Ultrasonic Sensor:** It is a device which uses to detect and sense ultrasonic sound. It is used in voice monitoring system where voice is lies within the ultrasonic range.

22. **Soil Moisture Sensor:** It is one kind of volumetric device which measures the amount of water in soil or presence of water in soil. It is used for roof top gardening or any kind of gardening purpose.

23. **Proximity Sensor:** It is used to detect the presence of any object by receiving a signal from it. It emits electromagnetic beam constantly and measures the radiation of reflected one. It is used for any presence of human being.

24. **Video Door Bell Theft Deterrent Sensor:** It's application quite similar with PIR sensor which detects the movement of human being. It is a combination of face recognition with alarm system.

25. **Microwave Sensor:** It is similar to Radar which emits microwave radiation and also receives which reflected back from the object's surface. It can detect any movement of object as well as senses other parameters like surface temperature of room.

26. **Hue Motion Sensor:** It can detect object movement and control the light on-off system.

27. **Ultrasonic Transducer:** It emits ultrasonic waves constantly and measures the reflected signal which etched by the object. It can convert the reflected signal into sound or vibrated energy.

28. **Glass Break Detector:** It detects any kind of sound noise or measures vibrations coming from any glass. It is used in bathroom or kitchen or room to secure the window of home.

29. **Force Sensor:** It converts a mechanical energy into electrical energy where the mechanical energy is depends on size, shape, and capacity of cells of the device. It is used to convert foot pressure into light energy and use at staircases, front door, etc.

30. **Piezo Film Sensor:** It just similar to Force sensor but here we use piezoelectric cells. Here the number of Piezo Film cells equivalent to generated mechanical energy.

31. **Position Sensor:** It detects the position in terms of angular and relative movement of the object. It is used in automated home cleansing system.

32. **Thermopile Sensor:** It is one kind of thermocouple sensor which measures temperature based on the distance of an object through its IR energy.

33. **Vibration Sensor:** It just like Glass break detector sensor which measures vibration.

34. **Magneto Resistive Sensor:** It measures the changes of radiation of magnetic field which created by an object and alerts from earthquake like natural calamities or some accident.

35. **Metal Detector Sensor:** It detects presence of any metal or ammunition and used as security system at front door.

36. **Solar Cell Light Sensor:** It is used to convert the solar energy into light energy.

37. **Flex Sensor:** It calculates the amount of bending or flexing of a surface. It is a collection of variable resistors which used in automated furniture cleaning system.

38. **Pulse Sensor:** It is as similar as heartbeat sensor which is a plug-&-play device to measure live heart pulse and attach with bed or chair of old people to monitoring their heartbeat rate.

39. **Geomagnetic Sensor:** It is a hardware-based device which measures the change of the earth's magnetic field to give stability of house with respect to natural calamities. It is collection of gravity sensor, orientation sensor, etc.

6.3   ADVANTAGE OF HOME AUTOMATION SYSTEM

Home is an example of the smallest unit of close system which is a collection of several integrated embedded systems. In today's world with modern trends of advanced technologies like IoT has a big impact on home automation system. All the home appliances are connected with each other through a set of sensors where they are passing various kinds of parameters with a lot of information such as device id, device name, device's time, device's set of configurations, device's status, etc. Nowadays, we know that non-renewable energies are too much precious, so we use home automation systems to utilize the energy and save our world. In case of home automation system, we consume lots of energy for controlling light, fan, air-conditioner, refrigerator, air-cooler, geyser, Microwave oven, Cooking Gas, Washing Machine, Floor Cleaner or Vacuum Cleaner, Dish Washer, Coffee Maker, Blender, Mixer, Toaster, Television, and many more where we utilize lots of electricity or sources of non-renewable energy like coal petroleum assets. To avoid the wastage of these kinds of resources we either move to solar cell technology or smart home automation system. There is various problematic phase for solar cell installation because variations of weather in tropical country. So, we move towards smart home system or home automation system.

It is used to save energy because we can control all kinds of switches very easily through one press mode. The fans, lights, and other appliances on-off activities happen very efficiently. It is very much easy to operate for old age people, as they can manage all devices from a single edge of point. There is no necessary to move throughout the floor or whole passage to control all home appliances. If they import some new high-end technological device then it is operated through the same process. As all the devices are connected through the internet so all devices are updated automatically without intervention of human operation.

For home we always prefer the security aspects and it is more needed for old people. In home automation system security patches are very crucial and well maintain due to the impact of cloud computing. The federated architecture of cloud computing enhances security prospect in smart home automation system. After COVID-19 pandemic situation, the number of cases regarding house-theft issues increases due to the poor financial conditions of ordinary people. To secure our home, we need to use such an authentic system with our front door of house which is based on either face recognition based or human monitoring system that can monitor every locking system from inside of the home. This kind of security management system is very much essential for those old aged people who are staying or spending a lot of time alone in the house. Smart home automation system when connects image recognition it creates a face authentic security system which uses at front door. As per the proper recognition of face of family members, the door lights automatically on or off. Through proximity sensor if any unknown person moves in front of front door, then light on at once and recognizes face. If it is known, then either open automatically or gives notifications to the inside with the name. So, there is a knowledge base (KB) which stores the image of each face of members with their name of all family members and relatives. There is another procedure may be introduced where the proximity sensor sense and opens camera then take a snap of that incoming person and shows in a monitor which is inside of the home and waiting for notifications or commands of the house member whether the locks open or not. In this scenario we can manage the security of home automation system. Not only that we can manage a wireless fidelity (Wi-Fi) concept by which we can share the device's password among the members of the house. Using this password, they can lock or unlock the front door, because security factor is one of the most important issues for home automation system. This kind of security factor is also working and save the house from the neighbors, thief, etc., through one fingertip when there is no member in the house due to spending of some vacation period.

For using floor management system, it is one of the most important pieces of equipment under this home automation system where not necessary to move the whole floor area physically to clean the entire house. It does automatically by the voice commands. So, it is not required whether the members of the house within the house or not to manage the working procedure of smart automatic floor cleaner. This is one of the examples by which we can save our effort and time using smart home automation system.

There is a concept of smart mirror as one of the components of smart home automation system where the mirror acts as a computer monitor or surveillance camera as per the voice commands. It can record all the incidents throughout the day which were happened in front of this and recorded everything. So, parents who are busy in day time due to office can monitor the house servant and how they can manage and gives services to their children in absence of them in house. Here this kind of smart mirror also acts as a house manager. It can show the status of incoming parcel which ordered through online, location of cabs, important news highlights, status of all house appliances, on its screen. It also finds the mobile which is in silent mode through the voice commands by enabling its default ringtone.

In smart home automation system, light management is one of the important factors for which people use it too much. There are no requirements of physically movement for switch on or off of the lights even with the help of object detection the light of staircases, corridor, balcony, and dining room automatically can manage which saves the lifetime of LED lights.

It maintains equilibrium of Oxygen and Carbon-dioxide within the home and gives a healthy hygiene environment. It calculates heartbeat of house members automatically, count of footsteps, oxygen capacity of rooms as well as ratios of Oxygen and Carbon-dioxide for each room. It checks the ratio each and every time and when it crosses the threshold value the windows automatically opens or exhaust fan automatically turns on. Not only windows and doors connect with it, the chairs, tables, bed can also connect with it. When the house members get relaxed and spend leisure time by sitting on chair or by lying on bed or eating foods on table at that time the sensors which are connect with this furniture can fetch the pulse rate and gives the status of your body through voice response by measuring the heart related health parameters just like a smartwatch.

So, this smart home automation system provides six sigma factors: control, savings, security, convenient, healthy, and comfort.

6.4 PROCESS FLOW

In this section, we first explain the Google assistance associated with mobile phone and it's working, as shown in Figure 6.1. After receiving instruction from devices, Google cloud console decides to work for it and sends the instruction to IFTTT. It detects the authentication and passes to mosquito server. Therefore, the device is recognized, and an update is made. A corresponding flow diagram is given in Figure 6.2.

6.5 RESULTS

The result shows that equipment can be switched on if the end-user asks to turn on and also it can be turned off. Moreover, it can search the web for the specific query and read out the results or inform user about the

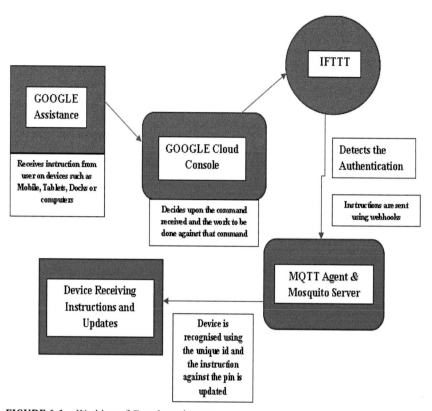

FIGURE 6.1   Working of Google assistance.

weather when queered for it. It also smartly helps the user to reduce the energy consumption by switching off the device when not needed. Figure 6.3 shows the complete circuit developed so far, along with the webcam which is used to take voice input from the user.

At this moment, the function of the relay becomes extremely important. Figure 6.4 shows the relay circuit at different working conditions, both high and low.

Corresponding variation of data in LDR is shown in Figures 6.5(a) and 6.5(b), respectively. At different times of day, the variation of LDR is given. It will be different at different days. The live reading of LDR data in mobile app along with the console is presented in Figure 6.6. Intensity variation with unit time for LDR is represented in Figure 6.7.

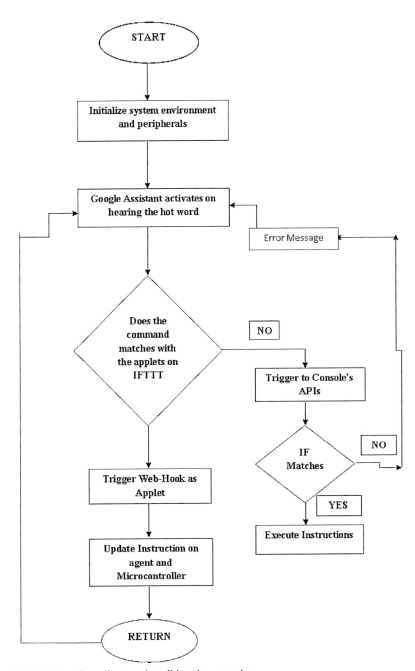

FIGURE 6.2   Flow-diagram describing the operation.

FIGURE 6.3   (a) Total circuit for operation; and (b) webcam is used to take voice input.

FIGURE 6.4   (a) Relay circuit at working condition; and (b) relay circuit when LDR value is low.

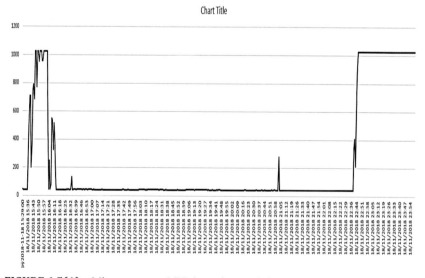

FIGURE 6.5(A)   Minute average LDR intensity graph from day '*1*' data sheet readings.

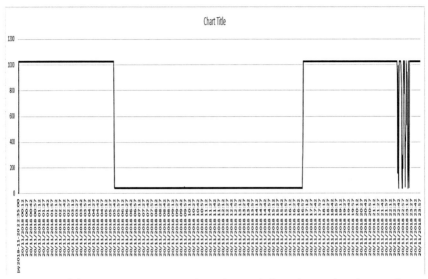

FIGURE 6.5(B)   Minute average LDR intensity graph from day '*n*' data sheet readings.

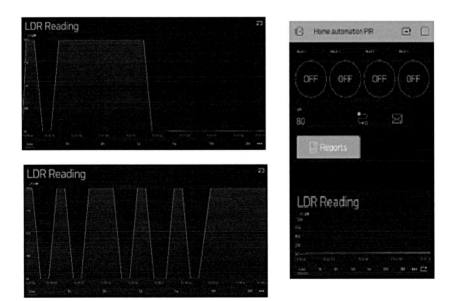

FIGURE 6.6   Live streaming results of the LDR sensor.

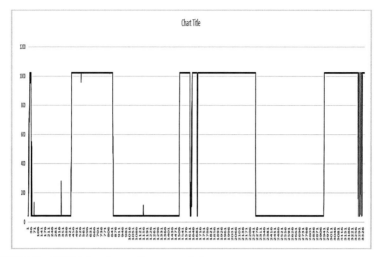

FIGURE 6.7   LDR intensity reading per unit time.

6.6   CONCLUSION

The aim of this chapter is to propose a cost-effective voice controlled (Google Assistant) home automation controlling general appliances found in one's home with intelligence that can be integrated within their existing circuit. The approach discussed in the chapter is already implemented, and hence may be termed as successful so far. This system is highly reliable and efficient for aged people and differently able person on a wheelchair who cannot reach the switch for the switching ON/OFF the device and are dependent on others. It is also applicable on the areas where user need appliance to work without reaching the switchboard and for saving energy. This system can be implemented within the existing switchboard by using a two-way switch.

KEYWORDS

- **Google assistance**
- **Google cloud console**
- **home automation**
- **motion sensing**
- **remote-controlled switching**
- **voice command**

REFERENCES

1. Parameshachari, B. D., Gopy, S. K., Hurry, G., & Gopaul, T. T., (2013). A study on smart home control system through speech. *International Journal of Computer Applications, 69*(19), 30–39.

2. Pal, S., Chauhan, A., & Gupta, S. K., (2019). Voice controlled smart home automation system. *International Journal of Recent Technology and Engineering, 8*(3), 4092–4093.

3. Rehman, A., Arif, R., & Khursheed, H., (2014). Voice controlled home automation system for the elderly or disabled people. *Journal of Applied Environmental and Biological Sciences, 4*(8S), 55–64.

4. Chin, H., Kim, J., Kim, I., Kwon, Y., Lee, K., & Yang, S., (2001). Realization of speech recognition using DSP (digital signal processor*). IEEE International Symposium on Industrial Electronics Proceedings,* 508–512.

5. Mewada, A., Mishra, A., Gupta, M., Dash, R., & Mulla, N., (2016). Voice controlled home automation. *International Journal of Advanced Research in Computer Science and Software Engineering, 6*(3), 161–164.

6. Nguyen, T. V., Lee, D. G., Seol, Y. H., Yu, M. H., & Choi, D., (2007). Ubiquitous access to home appliance control system using infrared ray and power line communication. In: *3rd IEEE/IFIP International Conference* (Vol 1, pp 1–4). Central Asia Tashkent, Uzbekistan.

7. Khiyal, M. S. H., Khan, A., & Shehzadi, E., (2009). SMS based wireless home appliance control system (HACS) for automating appliances and security. *Issue in Information Science and Information Technology, 6*, 887–894.

8. Jawarkar, N. P., Ahmed, V., & Thakare, R. D., (2007). Remote control using mobile through spoken commands. *IEEE International Consortium of Stem Cell Networks,* 622–625.

9. Mhadse, P. V., & Wani, A. C., (2013). Speaker identification based automation system through speech recognition. *International Journal of Emerging Technology and Advanced Engineering, 3*(1), 557–563.

10. Suralkar, S. R., Wani, A. C., & Mhadse, P. V., (2013). Speech recognized automation system using speaker identification through wireless communication. *IOSR Journal of Electronics and Communication Engineering, 6*(1), 11–18.

11. Santra, S., Mukherjee, P., & Deyasi, A., (2021). Cost-effective voice-controlled real-time smart informative interface design with google assistance technology. In: *Machine Learning Techniques and Analytics for Cloud Security.*

12. Sil, S., Daw, S., & Deyasi, A., (2021). Smart intelligent system design for accident prevention and theft protection of vehicle. *Lecture Notes in Electrical Engineering: 5th International Conference on Nanoelectronics, Circuits & Communication Systems* (pp. 523–530). Chapter 47.

13. Dhar, R., & Deyasi, A., (2020). Design of cost efficient LPG gas sensing prototype module embedded with accident prevention feature. In: *6th International Conference on Nanoelectronics, Computing & Communication Systems.*

CHAPTER 7

Applications of IoT and UAV in Smart Agriculture

CAPRIO MISTRY and ARIGHNA BASAK

Department of Electronics and Communication Engineering, Brainware University, Kolkata, West Bengal, India

ABSTRACT

This chapter demonstrates the numerous applications of intelligent agriculture, the advantages of IoT and UAV usage in agriculture, the variety of communication channels, and the connectivity issues and UAV and IoT restrictions in remote areas. The limitations of connection in terms of communication and transmission technologies were discussed in the chapter. Moreover, the chapter uses a system model to show how IOT and UAV are used to smart agriculture.

7.1 BACKGROUND

The population's reliance on the agriculture industry has grown dramatically. With the advent of technology, this epoch signifies a shift from traditional to the most imaginative techniques. Whatever people think about agricultural development, today's agriculture business is more precise, data-driven, and smarter than ever before. Almost every business, including "smart agriculture," has been transformed by the fast advancement of IoT-based technologies, which has shifted the sector from statistical to quantitative techniques. These fundamental alterations shock conventional agricultural systems and provide new chances for a variety of

System Design Using the Internet of Things with Deep Learning Applications.
Arpan Deyasi, Angsuman Sarkar, and Soumen Santra (Eds)

issues. Researchers, research institutions, academics, and the majority of nations throughout the world are driven to collaborate in order to examine the boundaries of this issue of masculinity development. The technology sector competes in order to provide better solutions. IoT, cloud control, big data analytics (BDA), and wireless sensor networks (WSNs), among other technologies, can give ample chance for situations to be forecasted, processed, and reviewed, and real-time activity to be restored. New fields in the field include the perception of heterogeneity and device interoperability using flexible, ascending, and powerful method models.

IoT may be applied in all aspects of life, including intelligent cities, intelligent homes, intelligent retail, and many more. The use of IoT in agricultural agriculture is critical, given the world population's ability to supply at much faster rates by 2050 to satisfy the highest peak of 9.6 billion. As a result, modern technologies, particularly IoT, are being used. IoT opens up the possibility of labor-free agriculture. It may also be used to support livestock, greenhouse farming, farm management, and other small-scale agricultural activities, in addition to large-scale agricultural operations. The most significant equipment used in IoT is sensors. Sensors are devices that collect vital data in order to perform the desired analysis. Sensors are commonly used in agriculture to collect data and determine levels of NPKs, pathogens, and humidity in soils. The internet of things (IoT) is being researched and explored for its potential applications in agriculture. Precision agriculture refers to smart agriculture that uses correct data to reach a conclusion. The sensors demonstrate a wide range of applications, issues, benefits, and drawbacks to help with IoT and farming.

As a result of the rapid technological advancement and decline in human capability, technology is becoming engaged in every step of our lives. Agriculture and irrigation are two areas where man's potential may be realized. Various sensors and electronic devices are employed in the industry to commercialize while keeping prices low in a few areas. To save money and improve agricultural workers' abilities, UAVs (unmanned aerial vehicles) can be used for reconnaissance, pesticide, and insecticide application, and bioprocessing fault detection. In this application, both single-mode and multi-mode UAV systems can be used. A single UAV system's skill can be surpassed by the proper collaboration and synchronization of a network of UAV clusters connected to terrestrial infrastructures, GCS, or satellites. As a result, the mobility model and requirements are the most effective routing protocol for each individual agricultural application.

7.2 INTRODUCTION

The Latin term 'Ager' means 'Land,' 'Culture' means 'Cultivation,' and 'Agricultura' means 'Agriculture.' This is one of the benchmarks for human growth. Ample revolutions were prepared throughout human history in order to make development on agricultural output with less properties and work. However, in all these ages, the high population level never allowed demand and supply to equal. Over 60% of the population is dependent on farming and around 12% of the total land area is subject to agricultural produce in accordance with the United Nations (UN) Food and Agriculture Organization (FAO) [1]. In 2050, the world's population is anticipated to reach 9.8 billion, an improvement of around 25% in relation to the existing state of affairs [2], according to the projection scenario. The population growth among the emerging nations is almost entirely projected [3]. On the other hand, the trend of urban expansion is expected, with nearly 70% of global population projections expected to continue at an accelerated pace through 2050 [4]. In India, the farming region occupies over 50% of the population, and the livelihood of around 61.5% of the Indian people depends mostly on agriculture [5, 6].

Over the last several decades, however, agricultural production is undergoing a fourth revolution through using ICT in conventional agriculture [7]. The majority of technologies are advantageous and may boost innovation in agriculture systems, such as BDA, and machine learning (ML), remote sensing, the IoT internet, UAVs [8, 9]. A wide range of agricultural constraints can be seen with smart farming in order to increase crop yields, ease costs and increase process inputs such as environmental conditions, soil development status, irrigation water, pesticides, and fertilizer products, the management of weeds, and the environment of greenhouse production [10]. Intelligent farming is a green technology practice since it reduces traditional agriculture's environmental footprint [9]. Precision agriculture may also reduce liquidation issues and productivity as well as impacts of climate change in the absence of any insignificant use of fertilizer and pesticides in agricultural crops [9–11]. IoT is one of the most recent advancements in wireless communication in radical technologies [12]. The fundamental concept is to link a range of physical devices to the internet with special addressing patterns. In a variety of vertical areas, such as transit, healthcare, industry, cars, smart home, and farming, IoT technology can be realistic [13]. IoT devices provide helpful information in an agricultural system on a variety of physical restrictions to optimize the farming process [10]. Wireless sensor networks (WSN) play an extremely reputable position in IoT technology, as wireless data transfer is a key aspect of IoT applications in many markets.

During the last two decades, the widespread usage of the internet has offered boundless advantages to organizations and people throughout the world. This achievement was mainly advantaged by the capability to offer customer and production services in real-time. The IoT has lately claimed to provide its revolutionary breakthroughs to the same advantages by altering its operational environment and raising consumer awareness and capabilities. In the areas of health, shopping, traffic, defense, intelligent homes, smart cities, and agriculture, IoT offers a number of solutions. In agriculture, IoT implementation is seen as the optimal solution, since continuous monitoring and control is essential in this sector. IoT is seen on a variety of occasions within the manufacturing chain of agriculture [14]. The framework built is even more interesting in dealing with node faults and rearranging the weak connectivity ties on the network on their own. In Ref. [15], an IoT management is proposed that tracks wind, soil, atmosphere, and water elements across a wide range of areas. In addition, IoT-based solutions for agricultural surveillance have been established based on their subdomains. Sub-domains listed include soil tracking, air surveillance, temperature monitoring, water monitoring, disease monitoring, location surveillance, environmental monitoring, insect monitoring, and fertilization surveillance [16, 17]. Precision farming, livestock, and greenhouses, which are clustered under various surveillance areas, are the major applications of IoT in agriculture. The wireless sensor network (WSN), which helps farmers collect relevant data via detectors, is used to track all these applications with different IoT-based sensors/devices. Certain IoT-based settings analyze and process remote data through cloud providers, helping scientists and farmers to decide better. Today, environmental control systems provide additional management and decision-making facilities through the development of current technologies. A custom landslide Risk Management System has been designed to enable fast deployments without user interference in hostile environments [18]. Data from various environmental parameters are furthermore conveyed to the user by warnings or by message to the authorities [19].

The objective of intelligent agriculture is to boost output, yield, and profitability by using many approaches, such as effective irrigation, the specific and exact application of pesticides and fertilizers on crops, and so on. Smart agriculture was made feasible by the developing IoT and UAVs. In order to ensure that information flows across different devices, such as sensors, relays, and gateways, IoT values its data via automated collection, translation, and access as a significant technology for intelligent agriculture. This enables smart farms cost-effective and timely production and management processes [20].

Furthermore, IoT reduces the impact of the intrinsic environment by allowing real-time response to infestations such as weeds, detection of diseases, weather surveillance and prevention, soils, etc. This facilitates the proper exploitation of commodities such as water, nutrients, or agrochemicals through UAV and IoT technologies. Moreover, these intelligent technologies have enhanced the quality of agricultural output and the environmental effects of agriculture. Because IoT and UAV-based intelligent agriculture have certain essential properties [21–23]:

i. **Field Monitoring:** Smart agriculture aims to reduce crop waste by improved surveillance, precise data collection and data analysis.

ii. **Intelligent Agricultural Activities:** These are designed to identify the position of animals in huge stables grazing in open spaces. Views and tracking: Technology also helps analyze the air quality, ventilation, and hazardous waste emissions in farms.

iii. **Greenhouse Applications:** Smart agriculture monitors microclimate conditions with the goal of improving fruit and vegetable production and quality in green houses.

iv. **Biomass Management:** Smart farming helps to regulate humidity and temperature in crops like grass as a measure for avoiding fungus and other microbial pollutants.

v. **Bioenergy Governance:** Smart farming is a measure of how to prevent fungus and other microbial pollutants in crops like straw and grass.

vi. **Intelligent Breeding:** It regulates the raising situations and well-being of offspring in animal farms.

In addition, the uses of UAVs are diversified, encompassing residential, military, trade, and government-related fields [13–18]. Examples include environmental control in the civil sector (e.g., pollution, plant health and industrial accidents). UAVs are being utilized to monitor and deliver information at post-catastrophe or disaster sites, and transfer medicines and other vital supplies in military and government zones. The delivery of commodities and products in both urban and rural regions is consumed by commercial uses. As part of IoT, UAVs, which therefore depend on sensors, embedded software, and antennas, get a two-way connectivity to remote control and monitoring applications [19]. The IoT produces a fast-paced cutting-edge environment where a huge number of clever things have the main notion to be orchestrated so that either directly by users or through specific software collects behavior and ideas are used and activated worldwide. With IoT, objects may develop active participants in everyday work through various communications technologies with rich and potential applications in the context of the "Intelligent City" vision [20]. It is predicted that around 30 billion distinct items will be members of this global community by 2022. It is expected. With the arrival of 5G UAV technology, these projections are projected to increase dramatically.

The aim of UAVs is to give a wider range of viewpoints in intelligent agriculture, for example, with an imagery analysis and agricultural surveillance [24]. Not only do UAVs increase picture analysis and farmland processing, but they also make it possible by patrolling the field of interest to thoroughly understand the situation [25]. UAVs can also be utilized to supply useful information through data transfer to grounded tracking stations. UAV usage is expanded in numerous agricultural fields, such as pesticide and fertile spraying, seed plantation, weakening of identity, fertility assessment, mapping, and planting. UAV is used in numerous fields. These recent advances in intelligent IoT and UAV-based agriculture help the world achieve the '2030 Sustainable Development Agenda' priorities, where the United Nations (UN) and international communities are setting the goal of eliminating hunger by 2030 [26].

7.3   UNMANNED AERIAL VEHICLE (UAV)

Unmanned aerial vehicles (UAVs) are a low-cost option for data analysis and sensing technologies in recent years. Fernosensing examines the properties of a targeted item from a distance through electromagnetic energy, and offers the benefits of integrity, non-invasiveness, timeliness, and adaptability. Far distant from actual data for remote sensing, agricultural productivity, and land [27–29], remote sensing determines soil parameters of the farmhouse. UAV can be chosen to obtain precise data according to the researchers in the field. It is regarded as a viable technology capable of producing high images (<1 m) in high spatial resolution and time, which is suited for prompt reactions in the generation of active crop and field information [30]. Less costly than traditional handled aircraft, the low altitude remote sensing (LARS), the tiny UAV may be one of the key reasons why UAV production exceeded market demand [31].

However, a UAV is a type of aircraft that functions without an on-board human pilot. Different types of UAVs for various purposes exist. Initially, the military used technology for the target practice, intelligence, and monitoring of certain enemy lands by the anti-aircraft group. However, the technology has gone beyond its basic objective and has become more important in many sectors of human effort in recent years [32]. Technological advances have allowed the increased adaption for numerous applications of unmanned air vehicles. UAVs can be utilized in cattle counting, animal observation, and eating behaviors and patterns linked to health from a cattle point of view. By using the information collected, farmers may give quick and efficient solutions to difficulties and challenges, make better management decisions, restore agricultural output, and finally generate higher incomes [33, 34].

UAV's advantages are to allow farmer's crops to hold pictures with a variety of camera filters to allow farms with multiple spectral imagery, to approve picture treatment and investigation, to provide better information about crop health while also recognizing crop areas that require particular forms of attention. The little UAV may be simply floated and maintained with minimal training, producing a fantastic alternative for farmers monitoring their crop expansion by combining farming with remote sensing information [33–35].

7.4 TYPES OF UAV

Many concepts may be utilized to obtain a categorization of UAVs [8], such as size, maximum take-offs, etc. These will also be taken into account in the examination of UAV flight regulation. The categorization has been established with the kind and amount of autonomy that would be most suitable in the field of agricultural missions [36–39].

Two primary sets may be detected depending on the kind of wing: rotating and fixing wings. Helicopters and multi-rooters are preparing for the first group (usually known as drones). The airflow consists of several rotors producing the appropriate lifting confidence. Their key benefit is the capability to accomplish aerial photography-enabled hovering flight since it enables the capture duration of cameras to be increased and therefore unfavorable light conditions reimbursed. The airflow also shows high performance at low speeds and is consistent with low-level flights with less danger. Due to its mechanical easiness in contrast to helicopters, which depend on a much more complex plate control mechanism, multi-rotor systems have become more common. Therefore, the multi-rotor flight can only be detected by changing the speed of numerous direct current (DC) engines, unlike helicopters, without a mechanically moving machines. Furthermore, a drop in prices for non-maintenance brushless engines and associated electronic supervisors has made it cost-effective for numerous civil applications responsibilities. As a result, a considerable number of drone companies for the provision of a large variety of systems came on the market. Commercial drones are mostly inadequate compared to helicopters with their lower payload capabilities. However, the number of rotors has risen from 4 to 6 or 8 to lessen that gap and drones with a maximum cargo capacity of 22 kg are exhibited on the market nowadays [40–43]. The cargo capacity not just increases when rotors are in use, but safety has developed similarly. Then, when one or more rotors are in failure, the aircraft are usually able to fly in a degraded state, allowing for safe arrivals. Conversely, fixed wing aircraft such as planes require airflow generation to increase aerodynamic surfaces (wing and aileron) by moving at high speed. The airplane hence cannot carry out static flights. Then, the speed cannot be reduced in the same manner as rotating wings, normally they are higher in height for safe flights. Furthermore, there is a great advantage over rotating wings to their maneuverability (i.e., quick rotations on the Vertical Axis). Due to regulatory requirements for flying UAVs, the maximum range for safe wing aircraft is larger than rotary wings and, in most scenarios, it does not produce any true benefits and yet most aircraft display enhanced load capacity. Consequently, the analysis concluded that rotary wings are more favorable than fixed wings to agricultural responsibility.

7.5   IOT FOR SMART AGRICULTURE

Intelligent farming is a current IoT-based agriculture idea that improves agricultural output. Farmers may employ smart agricultural products to successfully increase the quality and quantity of their crops by using fertilizers and other resources. On the farm 24 hours a day, farmers can't be modern. In addition, farmers could not employ various devices to measure the appropriate environmental conditions for their crops. IoT provides you an autonomous system that can operate in deprivation of any human guidance and can notify you to conclude correctly with regard to various problems they experience throughout agriculture. They can spread and tell the farmer even if the farmer is not on the ground, thereby enabling farmers to cultivate additional farmland. Universal population has been expected to reach Mark 9 billion by 2050 [1–5]. IoT applications must thus be made for farming so that big populations can sustain themselves and use farms and other belongings efficiently, since in certain locations, they are rare. Random weather conditions harm crops for global warming, and farmers are confronted with losses, thus IoT Smart Farming can take immediate steps to prevent this [23–25, 44, 45, 49–51].

However, several major apparatuses of IoT based Agriculture exist, like sensors, Internet access, wireless communications, data transmission, etc. Wireless communications have a significant component to play in effectively organizing IoT systems based on transmission distance, spectrum, and scenarios of application. The architecture of IoT is based primarily on 3 levels, as Figure 7.1 shows; specifically, the Physical layer for sensing, the network layer for data transmission contracts and the layer for application where data storage and data modification are observed [23].

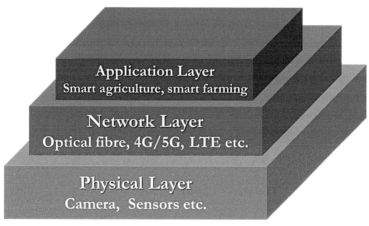

FIGURE 7.1   IoT architecture.

7.5.1 PHYSICAL LAYER

The hardware layer is comprised of many terminal devices, cameras, and sensors; WSNs; labels and readers for radio frequency identification (RFID). In this layer, wind speed, temperature, moisture, nutrition level, vehicle illnesses, insect infestations, etc., are utilized for assembling sensors. Composite information is controlled by the embedded devices and is uploaded for further processing and research onto a higher layer. These terminals and sensors are utilised for monitoring, monitoring and recognition of agriculture and livestock products. WSNs, for example, are commonly used for storage and logistics services management and monitoring [52]. Further, RFID technologies for networked devices are the most important pattern on the other hand. Electronic product code (EPC) data is stored using RFID tags which are read, turned on or controlled by RFID readers. Thus, in the agricultural arena, WSN technologies, RFID, and NFC are utilized to play an important role in item ID, tracking, monitoring, monitoring, and data storage on active or passive devices.

7.5.2 NETWORK LAYER

Sensors and devices on the network layer are required for connecting to the nearby nodes and the gateway to build a network. The sensor nodes at this layer co-operate and link to other gateways and nodes in a system to transmit data into a remote structure; for important information data are preserved, further examined, processed, and distributed [53].

7.5.3 APPLICATION LAYER

The uppermost level of the IoT architecture is the application layer in which IoT assists and benefits are external. In this layer, there are lots of brilliant phases or systems to monitor and monitor soil, water, and food, plants, and livestock. These layers support initial advice and organization of illnesses and insect pests, infestation, and agricultural management of safety; the result is improved manufacturing efficiency.

7.6 COMMUNICATION TECHNOLOGIES USED IN IOT-BASED SMART AGRICULTURE

Procedures for wireless communication have been set by wireless protocols and standards. IEEE 802.15.4, for example, is a wireless standard used to connect Internet access gates with the end nodes. In addition, Zigbee, Sigfox, EC-GSM, ONE-NET, and ISA100, 11a and LoRaWAN, DASH7, etc. [54] are further examples. The standards are grouped into transmission ranges given in Figure 7.2 [55].

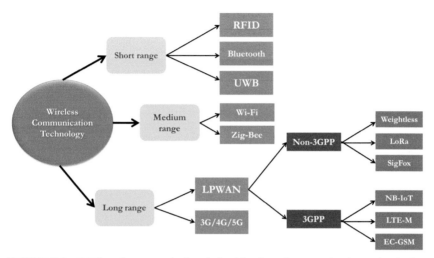

FIGURE 7.2   IoT-based smart agricultural classification of communication technologies.

IEEE 802.15.4-based protocols are the perfect choice [53] for wireless short distance applications, which is designed for broad-area networks with low energy (LPWANs). This data rate is from 20 kbps to 250 kbps and operates with a maximum outside viewing line (LoS) range of 100 m [56], and works with multiple frequency bands such as 433 MHz, 868 MHz, 91.5 and 2.4 GHz. For applications that are critical for medium-range communication the competing IEEE 802.11 standard are suitable. IEEE 802.11p is also an alternative to integrate high mobility conditions.

The wide range of communication, legal maximum transmission power of 1 W and the less disturbed 5.9 GHz ISM [50] appear to give this particular problem a great deal of attention in agricultural situations. Long distance communication technology is the only reliable and anticipated answer when it comes to coverage. The most suitable and reliable standard for precision farms in cellular communication technologies like as 3G, 4G, LTE, and 5G are large volumes of real-time data required for transmission and processing [57]. The 5G communication technology may also be used for delivering communication between devices in real-time and the device

(D2D) to enable for the positioning of vehicles. In addition, a huge number of equipment per square kilometer may be maintained [58]. Unlike LTE, 5G may be enabled on advanced frequency bands so that greater channel bandwidths may be used. 5G communication technologies, especially in rural regions, can enable the development of new capacity in agricultural equipment by providing greater data rates and broader transmission ranges according to the real-time connection model. The cellular network's obtainability and the economic potential of 5G technology in rural regions, however, are of debatable concern.

The most consistent of IoT future communication technologies are LoRa/LoRaWAN and IEEE 802.11ah. In addition, the major is an IEEE 802.11 family change, which was accessible in 2017 to provision IoT configurations like smart monitoring [50]. The situation usages 900 MHz bands and provides broader exposure and uses less than IEEE 802.15.4 energy. It connects 1,000 tactics across a one-kilometer radius with a single access point. LoRaWAN is another more favorable LPWAN standard for a battery-operated wireless node system.

Cellular communication approaches from 2G to 4G can be appropriate, dependent on the purpose and bandwidth necessity; however, the dependability, and even accessibility, of a cellular network in rural areas is a main apprehension. To overcome this, data transmission through satellite is another choice. However, the cost of this communication mechanism is high in this case is very high, which creates it not appropriate for small- and medium-sized farms. The superior of communication mode also be contingent on application requirements, like some farms require sensors that can activate with low data rate but necessary to work for long periods therefore requires long battery life. For such situations, a new choice of low power wide area network (LPWAN) is selected as a better resolution for cellular connectivity, not only in positions of long battery life but also in positions of a larger connectivity choice with reasonable rates [59]. Currently, the management of crops and pastures is one of the primary applications in which LPWAN networks are suitable for many additional agricultural uses.

In addition, a common communication label allows UAVs to connect with Ground Control Station for UAV farming (MAVLink) (GCS). It links up between the platform calculation, UAV monitoring platform and the GCS application platform [60]. MAVLink is used for transmitting directions, global satellite navigation system (GNSS) position and UAV speed.

The space of communication between the UAV and the GCS is subject to UAV circumstances. On the other hand, when the UAV is located inside the LOS it may link up to 2 km [61]. At present, UAVs are scheduled for re-start mode, so that when communication is disrupted, the UAVs may immediately return to its first location. This approach is used to stop undesirable incidents or UAV losses [24, 62, 63]. Also available for GCS and UAVs, such as ZigBee, radio-frequency modules, and other communicators, are a few different types of communications systems. By using other technologies, such as phone applications, the communication range may be increased. The arrival of 5G mobile technology, will be very useful to plot high definition in order to restore infrastructure and data processing rates [64, 65].

7.7   UAV FOR SMART AGRICULTURE

Nowadays, in various industries, including farming areas, the IoT system has made great growth. But communication services such as basic stations or Wi-Fi are highly inadequate when taking on agriculture which restrict the development of the IoT industry. Their arrangements for communication and related facilities in emerging countries and rural regions, one of the primary gaps in the IoT business, are significantly worse. In the lack of reliable communication arrangements, the data produced by the wireless sensors cannot be transferred. In such a case, UAVs are making an alternative method. In the course of data processing and inquiry, the UAV system interfaces with the wireless sensors over great distances. The UAV, also known as drones, was widely used amongst the most precious innovations in smart agriculture [25, 66]. Drones or UAVs are widely utilized by farmers for observation and control of farm growth. Some UAVs are sprayed, others are very insecticidal in the rugged terrain when human mobility is not possible and crops keep different heights.

In addition, UAVs, better known as drones, may be flown over thousands of hectares in the fields, fastened with high-resolution cameras and accurate sensors. In the fields of forestry and crop observation, in particular, the investigational function for wide zones in all agricultural applications is quite serious [67]. A rapid, low-cost, real, and comprehensive monitoring of accurate data acquisition and broadcasting is therefore essential for agriculture. Currently, aerial photos from a field or a farm zone are usually employed by two choices: satellite and aircrew. Both options are suitable for a scenic view, but when it comes to PC views, the options confront serious challenges with their qualities. These pictures are not respectable in resolution and do not give image quality that is essential throughout the inquiry. In addition, it is also difficult to acquire pictures regularly and the frequency of visits is difficult. In addition to this, it is another serious concern because controls outside the cloud are likely to be crowded in a harsh environment.

But UAV is an eye-in-heaven platform that resolves or eliminates the previously mentioned micro-vision difficulties. The photographs occupied by the UAV are generally much better than satellite photos in their worth depending on the included camera resolution and, in particular, regulate according to applications. UAVs maintain quicker, quicker and more advantaged NDVI to evaluate agricultural conditions such as weed mapping, leaf evaluations, etc., and to provide farmers with immediate reaction in good time. Depending on the frequency of the UAV, even if you wish it several times in one day, and if it is not rains it will alter weather conditions. UAVs are considered as the future of precision farming because of the advantages indicated.

Not only incalculability but also UAVs are required to deliver hefty payloads in specific applications of pesticide and fertilizer. This makes efficient battery utilization crucial for spreading flight duration in such conditions. Many characteristics may be examined to improve the efficiency of the drone in this resolution. At first, pick the proper circumstances for your flight, e.g., air or climatic direction. In the end, try to include and correctly put optimal payloads. In order to do this, the payload can be allocated near the field, improved in small numbers, and refilled again for heavy amounts. In addition, optimal path choice has a major significance, depending on area size and frequency of visitors. Many routing designs for this resolution, such as in Refs. [59, 60], are planned especially for UAVs, such that the choice and implementation of the proper arrangement can make significant changes. New activities such as the tethering system can assist the spraying of pesticide and UAV-based crops in which drones are important for flying heavy payloads. In UAV tying, the connector provides power via the long connection so that, as long as the farmers have electricity on the field, hefty batteries don't have to be lifted most of all. Currently, farming is regarded one of the most successful areas in which UAVs can offer several answers to a number of major and long-term challenges. Below are some major applications where drones already play key roles in support of farmers by means of the system concept.

7.8   APPLICATIONS OF IOT AND UAV IN SMART AGRICULTURE THROUGH SYSTEM MODEL

In numerous applications of intelligent agriculture, IoT, and UAVs may be used. This section discusses several IoT and UAV applications through system models in intelligent agriculture:

1. **Smart Crop Monitoring:** Crop surveillance indicates that particular agricultural restrictions are adequately sensed. One of the key elements of intelligent agriculture is automatic observation. Spontaneously positioned sensors can senses and transmit data to a gateway for further research and control. Cutting limitations including leaf area index [68], plant height, color, form, sheet sizes [69], etc., are monitored using sensors. IoT and UAV are also convenient in controlling soil moisture [70, 71]; agriculture water constraints like pH level [72]; and also, climate parameters like wind speed, Wind direction, precipitation, radiation, wind direction, air pressure, temperature, relative humidity, etc. [73, 74]. Remote sensing is also quite efficient in its presence. Due to the triviality of sensors, distant sensors are connected in lower UAV altitudes and may therefore efficiently and efficiently regulate crops. High resolution recordings are therefore accomplished by eliminating various forms of artificial conditions, such as weather.

2. **Smart Pest Management:** Pest control usually relies on three characteristics: sensing, assessing, and therapy. The state-of-the-art infections and pest assessment processing techniques are determined by employing UAVs or remote sensing satellites to create raw images across the crop zone. Remote sensor protection often provides wide ranges, consequently offering lesser cost for increased production. UAV IoT sensors, on the other hand, are employed for the maintenance in each angle of the crop series of other data assembly capabilities such as environmental sampling, plant health, and plague.

3. **Smart Irrigation:** Agricultural UAVs with cameras are capable of giving excellent perceptions of the particular areas affected. The farmer is able to control low-soil humidity areas, dehydrous

crops, and water-logged regions by means of cameras and, in general, is an understanding of the overall health status of the agricultural crops. Either traditional farming, inappropriately carried out or extremely costly, such accurate observation was not likely as professionals were recruited in order to carry out the work and achieve satisfying answers. At now, however, UAVs help farmers to arrange them themselves. They offer added benefits.

4. **Livestock Monitoring (Animal Husbandry):** Livestock monitoring is nowadays a very tough part of agriculture, which engage a very large number of labor efficiency, which cause drawback of cost-effectiveness. In this IoT enable 5G drone it is possible to take remotely surveillance on livestock monitoring. The interests of a growing number of scientists have drawn unmanned air vehicles (UAVs), as an agricultural tool. We examine in this research the challenge of deploying UAVs to track and monitor cattle such as bovine animals and ovine animals in pastures. We presume that all targeted animals have GPS collars and cannot disregard the movement of any targeted animal. In addition, the number of UAVs covering the whole pasture is assumed and we strive to discover the ideal deployment of UAVs to minimize the average distance between animals. First, we introduce a sweeping coverage technique using UAVs. With UAVs, the initial positions of all targeted animals may be gathered to cover the whole pasture. Then decide and update the deployment of the UAV by streaming a k-means clustering based on the original positions.

5. **Forecasting:** Projections are the primary characteristics of intelligent farming that give real-time data and previous information to estimate and compute certain critical parameters. Scientific demonstration and machine learning are samples of devices that work for prediction. UAV provides different machine learning models like the regression model for approximating phosphorus amount in the soil [48], forecasting moisture of soil or detection of plant disease [47], Artificial Neural Networks for forecasting temperatures of the field [46], etc.

7.8.1   SYSTEM MODEL

IoT is an emerging cutting-edge technology, which is dealing with different multidisciplinary Engineering frameworks for domestic or commercial Automation libraries. If we consider 3 Gateways as such gateway1, gateway2, gateway3, … If we follow OSI model over here network layers are to be more concentrated. Three gateways are enough to optimize the network traffic for user to UAV band and vice versa. After sensing data transmitted through gateway to online cloud server and from the client machine (mobile, tablet, computer) it may be accessed through different portal with real-time readings (Figure 7.3).

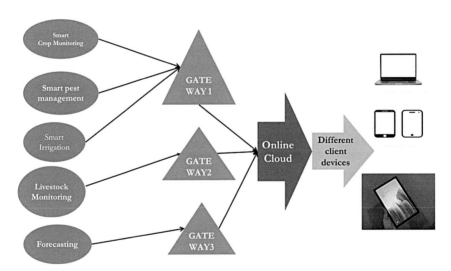

FIGURE 7.3   System model.

7.9   CONCLUSIONS

This chapter shows many uses of intelligent agriculture, benefits, and the use of IoT and UAV in agriculture, the diverse methods of communication, and acknowledges the connection challenges and limitations of IoT and UAVs in distant locations. The chapter described the limits on connection in terms of communication and transmission technologies. Moreover, the chapter demonstrates the application of IoT and UAV for smart agriculture through a system model.

KEYWORDS

- **big data analytics**
- **Food and Agriculture Organization**
- **Internet of Things**
- **machines learning**
- **unmanned aerial vehicles**
- **wireless sensors networks**

REFERENCES

1. Zavatta, G., (2014). *Agriculture Remains Central to the World Economy, 60% of the Population Depends on Agriculture for Survival.* NetExpo, [Online]
2. *World Population Projected to Reach 9.8 Billion in 2050, and 11.2 Billion in 2100* (2019). https://www.un.org/development/desa/en/news/population/world-population-prospects2017.html (accessed on 11 October 2022).
3. *How to Feed the World 2050,* (2013). Available: http://www.fao.org/leadmin/templates/wsfs/docs/expert_paper/How_to_Feed_the_World_in_2050.pdf (accessed on 11 October 2022).
4. Chouhan, S. S., Singh, U. P., & Jain, S., (2020). Applications of computer vision in plant pathology: A survey. *Arch. Comput. Methods Eng., 27*(2), 611–632.
5. Sawe, & Benjamin, E., (2017). WorldAtlas. *Countries Most Dependent on Agriculture,* Available: https://www.worldatlas.com/articles/countriesmost-dependent-onagriculture.html (accessed on 11 October 2022).
6. FAO in India, Food and Agricultural Organization of United Nations, (2017). Accessed, Available: http://www.fao.org/india/fao-inindia/india-at-a-glance/en/2017 (accessed on 11 October 2022).
7. Sundmaeker, H., Verdouw, C., Wolfert, S., & Prez, F. L., (2016). Internet of food and farm 2020. *Digitizing the Industry - Internet of Things Connecting Physical, Digital and Virtual Worlds, 2,* 129–151.
8. Wolfert, S., Ge, L., Verdouw, C., & Bogaardt, M. J., (2017). Big data in smart farming a review. *Agricultural Systems, 153,* 69–80.
9. Walter, A., Finger, R., Huber, R., & Buchmann, N. (2017). Opinion: Smart farming is key to developing sustainable agriculture. *Proceedings of the National Academy of Sciences, 114*(24), 6148–6150.
10. Nukala, R., Panduru, K., Shields, A., Riordan, D., Doody, P., & Walsh, J., (2016). Internet of things: A review from farm to fork. In: *2016 27th Irish Signals and Systems Conference (ISSC)* (pp. 1–6).

11. Wong, S., (2019). *Decentralized, Off-Grid Solar Pump Irrigation Systems in Developing Countries – Are They Pro-Poor, Pro-Environment and Pro-Women?* (pp. 367–382). Springer International Publishing, Cham.

12. Atzori, L., Iera, A., & Morabito, G., (2010). The internet of things: A survey. *Computer Networks, 54*(15), 2787–2805.

13. Al-Fuqaha, A., Guizani, M., Mohammadi, M., Aledhari, M., & Ayyash, M., (2015). Internet of things: A survey on enabling technologies, protocols, and applications. *IEEE Communications Surveys Tutorials, 17*(4), 2347–2376.

14. Medela, A., Cendón, B., González, L., Crespo, R., & Nevares, I., (2013). IoT multiplatform networking to monitor and control wineries and vineyards. In: *Proceedings of the 2013 Future Network Mobile Summit* (pp. 1–10). Lisboa, Portugal.

15. Zheng, R., Zhang, T., Liu, Z., & Wang, H., (2016). An EIoT system designed for ecological and environmental management of the xianghe segment of China's grand canal. *Int. J. Sustain. Dev. World Ecol., 23*, 372–380.

16. Torres-Ruiz, M., Juárez-Hipólito, J. H., Lytras, M. D., & Moreno-Ibarra, M., (2016). Environmental noise sensing approach based on volunteered geographic information and Spatio-temporal analysis with machine learning. In: *Proceedings of the International Conference on Computational Science and Its Applications* (pp. 95–110). Beijing, China.

17. Hachem, S., Mallet, V., Ventura, R., Pathak, A., Issarny, V., Raverdy, P. G., & Bhatia, R., (2015). Monitoring noise pollution using the urban civics middleware. In: *Proceedings of the 2015 IEEE First International Conference on Big Data Computing Service and Applications* (pp. 52–61). Redwood City, CA, USA.

18. Giorgetti, A., Lucchi, M., Tavelli, E., Barla, M., Gigli, G., Casagli, N., & Dardari, D., (2016). A robust wireless sensor network for landslide risk analysis: System design, deployment, and field testing. *IEEE Sens. J., 16*, 6374–6386.

19. Liu, Z., Huang, J., Wang, Q., Wang, Y., & Fu, J., (2013). Real-time barrier lakes monitoring and warning system based on wireless sensor network. In: *Proceedings of the 2013 Fourth International Conference on Intelligent Control and Information Processing (ICICIP)* (pp. 551–554). Beijing, China.

20. Glaroudis, D., Iossifides, A., & Chatzimisios, P., (2020). Survey, comparison and research challenges of IoT application protocols for smart farming. *Comput. Netw., 168*, 107037.

21. Osuch, A., Przygodzinski, P., Rybacki, P., Osuch, E., Kowalik, I., Piechnik, L., Przygodzinski, A., & Herkowiak, M., (2020). *Analysis of the Effectiveness of Shielded Band Spraying in Weed Control in Field Crops, 10*, 475.

22. Panchasara, H., Samrat, N., & Islam, N., (2021). Greenhouse gas emissions trends and mitigation measures in Australian agriculture sector: A review. *Agriculture, 11*, 85.

23. Villa-Henriksen, A., Edwards, G. T., Pesonen, L. A., Green, O., & Sørensen, C. A. G., (2020). Internet of things in arable farming: Implementation, applications, challenges and potential. *Biosyst. Eng., 191*, 60–84.

24. Kim, J., Kim, S., Ju, C., & Son, H. I., (2019). Unmanned aerial vehicles in agriculture: A review of perspective of platform, control, and applications. *IEEE Access, 7*, 105100–105115.

25. Mogili, U. R., & Deepak, B., (2018). Review on application of drone systems in precision agriculture. *Procedia Comput. Sci., 133*, 502–509.

26. Gubbi, J., Buyya, R., Marusic, S., & Palaniswami, M., (2013). Internet of things (IoT): A vision, architectural elements, and future directions. *Future Gener. Comput. Syst., 29*, 1645–1660.

27. Berni, J., Zarco-Tejada, P., Suárez, L., González-Dugo, V., & Fereres, E., (2009). Remote sensing of vegetation from UAV platforms using lightweight multispectral and thermal imaging sensors. *Int. Arch. Photogramm. Remote Sen. Spatial Inform Sci, 38*(6).

28. Elsenbeiss, H., & Sauerbier, M., (2011). Investigation of UAV systems and flight modes for photogrammetric applications. *The Photogrammetric Record,* 400–421.

29. Thomasson, J. A., Sui, R., & Ge, Y., (2011). Remote sensing of soil properties in precision agriculture: A review. *Front Earth Sci.,* 229–238.

30. Elarab, M., Ticlavilca, A. M., Torres-Rua, A. F., Maslova, I., & McKee, M., (2015). Estimating chlorophyll with thermal and broadband multispectral high resolution imagery from an unmanned aerial system using relevance vector machines for precision agriculture. *International Journal of Applied Earth Observation and Geoinformation,* 32–42.

31. Zhang, Y., Wang, L., & Duan, Y., (2016). Agricultural information dissemination using ICTs: A review and analysis of information dissemination models in China. *Information Processing in Agriculture,* 17–29.

32. Cano, E., Horton, R., Liljegren, C., & Bulanon, D., (2017). Comparison of small unmanned aerial vehicles performance using image processing. *Journal of Imaging,* 1–14.

33. Mukherjee, A., Misra, S., Sukrutha, A., & Raghuwanshi, N. S., (2020). Distributed aerial processing for IoT-based edge UAV swarms in smart farming. *Comput. Netw.,* 167

34. Bacco, M., Berton, A., Gotta, A., & Caviglione, L., (2018). IEEE 802.15.4 air-ground UAV communications in smart farming scenarios. *IEEE Commun. Lett., 22*(9), 1910–1913,

35. Allred, B., Martinez, L., Fessehazion, M. K., Rouse, G., Williamson, T. N., Wishart, D., Koganti, T., et al., (2020). Overall results and key findings on the use of UAV visible-color, multispectral, and thermal infrared imagery to map agricultural drainage pipes. *Agricult. Water Manage., 232,* 106036.

36. Vroegindeweij, B. A., Van, W. S. W., & Van, H. E., (2014). Autonomous unmanned aerial vehicles for agricultural applications. In: *Proceeding. International Conference of Agricultural Engineering (AgEng),* 8.

37. Sylvester, G., (2018). *E-Agriculture in Action: Drones for Agriculture.* Bangkok: Published by Food and Agriculture Organization of the United Nations and International Telecommunication Union;

38. Chapman, S., Merz, T., Chan, A., Jackway, P., Hrabar, S., Dreccer, M., et al., (2014). Pheno-copter: A low-altitude, autonomous remote-sensing robotic helicopter for high-throughput field based phenotyping. *Agronomy, 4*(2), 279–301.

39. Sugiura, R., Noguchi, N., & Ishii, K., (2005). Remote-sensing technology for vegetation monitoring using an unmanned helicopter. *Biosystems Engineering, 90*(4), 369–379.

40. Torres-Sanchez, J., Lopez-Granados, F., De Castro, A., & Pena-Barragan, J., (2013). Configuration and specifications of an unmanned aerial vehicle (UAV) for early site specific weed management. *PLoS One, 8*(3), 58210.

41. Shafian, S., Rajan, N., Schnell, R., Bagavathiannan, M., Valasek, J., Shi, Y., et al., (2018). Unmanned aerial systems-based remote sensing for monitoring sorghum growth and development. *PLoS One, 13*(5), 0196605.

42. Zhang, J., Basso, B., Price, R. F., Putman, G., & Shuai, G., (2018). Estimating plant distance in maize using unmanned aerial vehicle (UAV). *PLoS One, 13*(4), 0195223.

43. Lv, M., Xiao, S., Tang, Y., & He, Y., (2019). Influence of UAV flight speed on droplet deposition characteristics with the application of infrared thermal imaging. *International Journal of Agricultural and Biological Engineering, 12*(3), 10–17.

44. Ayaz, M., Ammad-Uddin, M., Sharif, Z., Mansour, A., & Aggoune, E. H. M., (2019). Internet-of-things (IoT)-based smart agriculture: Toward making the fields talk. *IEEE Access, 7*, 129551–129583.

45. Osuch, A., Przygodzinski, P., Rybacki, P., Osuch, E., Kowalik, I., Piechnik, L., Przygodzinski, A., & Herkowiak, M., (2020). Analysis of the effectiveness of shielded band spraying in weed control in field crops. *Agronomy, 10*, 475.

46. Aliev, K., Pasero, E., Jawaid, M. M., Narejo, S., & Pulatov, A., (2018). Internet of plants application for smart agriculture. *Int. J. Adv. Comput. Sci. Appl., 9*, 421–429.

47. Gao, J., Nuyttens, D., Lootens, P., He, Y., & Pieters, J. G., (2018). Recognizing weeds in a maize crop using a random forest machine-learning algorithm and near-infrared snapshot mosaic hyperspectral imagery. *Biosyst. Eng., 170*, 39–50.

48. Estrada-López, J. J., Castillo-Atoche, A. A., Vázquez-Castillo, J., & Sánchez-Sinencio, E., (2018). Smart soil parameters estimation system using an autonomous wireless sensor network with dynamic power management strategy. *IEEE Sens. J. 18*, 8913–8923.

49. Agarwal, P., Singh, V., Saini, G., & Panwar, D., (2019). sustainable smart-farming framework: Smart farming. In: *Smart Farming Technologies for Sustainable Agricultural Development* (pp. 147–173). IGI Global: Hershey, PA, USA.

50. Bacco, M., Berton, A., Ferro, E., Gennaro, C., Gotta, A., Matteoli, S., Paonessa, F., et al., (2018). Smart farming: Opportunities, challenges and technology enablers. In: *Proceedings of the 2018 IoT Vertical and Topical Summit on Agriculture-Tuscany (IOT Tuscany)* (pp. 1–6). Tuscany, Italy.

51. Hunter, M. C., Smith, R. G., Schipanski, M. E., Atwood, L. W., & Mortensen, D. A., (2017). Agriculture in 2050: Recalibrating targets for sustainable intensification. *Bioscience, 67*, 386–391.

52. Tzounis, A., Katsoulas, N., Bartzanas, T., & Kittas, C., (2017). Internet of things in agriculture, recent advances and future challenges. *Biosyst. Eng., 164*, 31–48.

53. Gubbi, J., Buyya, R., Marusic, S., & Palaniswami, M., (2013). Internet of things (IoT): A vision, architectural elements, and future directions. *Future Gener. Comput. Syst. 29*, 1645–1660.

54. Suhonen, J., Kohvakka, M., Kaseva, V., Hämäläinen, T. D., & Hännikäinen, M., (2012). *Low-Power Wireless Sensor Networks: Protocols, Services and Applications*. Springer Science & Business Media: New York, NY, USA,

55. Feng, X., Yan, F., & Liu, X., (2019). Study of wireless communication technologies on the internet of things for precision agriculture. *Wirel. Pers. Commun., 108,* 1785–1802.

56. Jain, R., (2016). *Wireless Protocols for IoT Part II: IEEE 802.15.4Wireless Personal Area Networks.* IEEE: Saint Louise, MO, USA,

57. Shi, X., An, X., Zhao, Q., Liu, H., Xia, L., Sun, X., & Guo, Y., (2019). State-of-the-art internet of things in protected agriculture. *Sensor, 19,* 1833.

58. Le, N. T., Hossain, M. A., Islam, A., Kim, D. Y., Choi, Y. J., & Jang, Y. M., (2016). Survey of promising technologies for 5G networks. *Mob. Inf. Syst.*

59. Beecham Research, (2016). *An Introduction to LPWA Public Service Categories: Matching Services to IoT Applications.*

60. Atoev, S., Kwon, K. R., Lee, S. H., & Moon, K. S., (2017). Data analysis of the MAVLink communication protocol. In: *Proceedings of the2017 International Conference on Information Science and Communications Technologies (ICISCT)* (pp. 1–3). Tashkent, Uzbekistan.

61. Salaan, C. J., Tadakuma, K., Okada, Y., Sakai, Y., Ohno, K., & Tadokoro, S., (2019). Development and experimental validation of aerial vehicle with passive rotating shell on each rotor. *IEEE Robot. Autom. Lett., 4,* 2568–2575.

62. Coombes, M., Chen, W. H., & Liu, C., (2018). Fixed wing UAV survey coverage path planning in wind for improving existing ground control station software. In: *Proceedings of the 2018 37th Chinese Control Conference (CCC)* (pp. 9820–9825). Wuhan, China.

63. Bhandari, S., Raheja, A., Chaichi, M. R., Green, R. L., Do, D., Pham, F. H., Ansari, M., et al., (2018). Lessons learned from UAV-based remote sensing for precision agriculture. In: *Proceedings of the 2018 International Conference on Unmanned Aircraft Systems (ICUAS)* (pp. 458–467). Dallas, TX, USA.

64. Alsalam, B. H. Y., Morton, K., Campbell, D., & Gonzalez, F., (2017). Autonomous UAV with vision based on-board decision making for remote sensing and precision agriculture. In: *Proceedings of the 2017 IEEE Aerospace Conference* (pp. 1–12). Big Sky, MT, USA.

65. Saha, A. K., Saha, J., Ray, R., Sircar, S., Dutta, S., Chattopadhyay, S. P., & Saha, H. N., (2018). IOT-based drone for improvement of crop quality in agricultural field. In: *Proceedings of the 2018 IEEE 8th Annual Computing and Communication Workshop and Conference (CCWC)* (pp. 612–615). Las Vegas, NV, USA.

66. Muchiri, N., & Kimathi, S., (2016). A review of applications and potential applications of UAV. In: *Proceedings of the Sustainable Research and Innovation Conference* (pp. 280–283). Rovinj, Croatia.

67. Uddin, M. A., Mansour, A., Le Jeune, D., Ayaz, M., & Aggoune, E. H., (2018). UAV-assisted dynamic clustering of wireless sensor networks for crop health monitoring, *Sensors, 18*(2), 555.

68. Orlando, F., Movedi, E., Coduto, D., Parisi, S., Brancadoro, L., Pagani, V., Guarneri, T., & Confalonieri, R., (2016). Estimating leaf area index (LAI) in vineyards using the pocket LAI smart-app. *Sensors, 16,* 2004.

69. Dadshani, S., Kurakin, A., & Amanov, S., (2015). Non-invasive assessment of leaf water status using a dual-mode microwave resonator. *Plant Methods, 11,* 8.

70. Sahota, H., Kumar, R., & Kamal, A., (2011). A wireless sensor network for precision agriculture and its performance. *Wirel. Commun. Mob. Comput., 11*, 1628–1645.

71. Vellidis, G., Garrick, V., Pocknee, S., Perry, C., Kvien, C., & Tucker, M., (2007). How wireless will change agriculture. In: *Proceedings of the Sixth European Conference on Precision Agriculture (6ECPA)* (pp. 57–67). Skiathos, Greece.

72. Islam, N., Ray, B., & Pasandideh, F., (2020). IoT based smart farming: Are the LPWAN technologies suitable for remote communication? In: *Proceedings of the 2020 IEEE International Conference on Smart Internet of Things (SmartIoT)* (pp. 270–276). Beijing, China.

73. Crabit, A., Colin, F., Bailly, J. S., Ayroles, H., & Garnier, F., (2011). Soft water level sensors for characterizing the hydrological behavior of agricultural catchments. *Sensors, 11*, 4656–4673.

74. Navulur, S., & Prasad, M. G., (2017). Agricultural management through wireless sensors and internet of things. *Int. J. Electr. Comput. Eng., 7*, 3492.

CHAPTER 8

A Study on Data Security in Cloud Computing: Traditional Cryptography to the Quantum Age Cryptography

SRINIDHI KULKARNI,[1] RISHABH KUMAR TRIPATHI,[1] and MEERA JOSHI[2]

[1]*Department of Computer Science & Engineering and Information Technology, International Institute of Information Technology, Bhubaneswar, Odisha, India*

[2]*Department of Mathematics, Aurora's Degree and PG College, Hyderabad, Telangana, India*

ABSTRACT

The rapidly advancing needs for scalable technologies which accommodate and handle the huge data inspired the idea of the Cloud. In Cloud systems, the services, platforms, or infrastructure are provided to the client as a service based on the requirements specified. The main element that connects the client and the virtual cloud is the data which is to be saved in or accessed from the cloud. The major concern around this data flow and cloud interaction is the security, privacy of the data which is in transit or at rest. The traditional approaches to secure a system are through the enforcement of the cryptographic algorithms, whose backbone is the complex mathematics and the functions which are computationally costly to decode. But as we advance as a modern generation of technology with upgraded computing capabilities such as quantum computing, the idea of

System Design Using the Internet of Things with Deep Learning Applications.
Arpan Deyasi, Angsuman Sarkar, and Soumen Santra (Eds)
© 2024 Apple Academic Press, Inc. Co-published with CRC Press (Taylor & Francis)

securing systems on the basis of computationally hard problems is not completely secure as quantum computation emerges as a threat to the pure cryptosystems. This prompts to the necessity of developing hybrid cryptosystems, which are more secure than the pure cryptosystems. In this chapter, we present a comparative analysis between various traditional cryptosystems and hybrid cryptosystems considering the factors such as the length of the key or cipher and the time for encryption and decryption. In this chapter, we also present a study of various security enhancing concepts, such as (i) machine learning and deep learning-based methods; (ii) quantum cryptography methods; and (iii) post-quantum cryptography concepts, which secure the cloud ecosystem despite the threats posed due to emergence of quantum computers or supercomputers. This chapter aims to contribute to the domain knowledge of cloud security and the different ways to enhance it.

8.1 INTRODUCTION

Cloud computing is a newly emerged technology with extensive scope of advancement. Initially, the time-sharing concept-based data centers assigned the tasks (Jobs) to the mainframe of the companies. During this period, the Jobs were submitted by the user to the operator directly. As the technology advanced, the companies began using VPN-based networks (virtual private network). The general notion was thought the demarcation in such systems between the end users and the servers was a cloud. With experimentations and using more optimized algorithms, the computing power and scalability of this cloud was further expanded, this was the beginning of the cloud computing revolution. The main notion of introducing this technology was the competent usage of the resources for the end user even at a lower bandwidth [1–4]. Many believe that the idea of providing computation as a utility was for the first time proposed by scientist John McCarthy [5] who introduced the novel idea of renting of the computing resources, which are time shared services, to the companies which were financially incapable of affording these technologies. Cloud mainly offers the service of storing and providing on-demand access to data over the internet. The utility rate of cloud services is expanding exponentially as the contemporary paradigm of technology is about data revolution. In the IT industry, the term was first popularized by Amazon releasing their product "Elastic Compute Cloud" in 2006 which changed the perspective of "Providing a Service." Cloud Services provide on demand the infrastructure, platform, and software as a service (SaaS). Idea of Cloud Computing is inspired from the Client-Server model as even in cloud architectures there is a client, who requests for the services and a Server (cloud service provider) who processes the requests. Figure 8.1 depicts the high-level graphical view of cloud computing architectures.

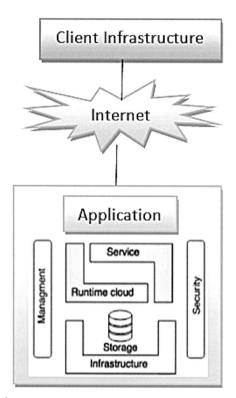

FIGURE 8.1   Cloud ecosystem.

8.1   CHARACTERISTICS OF CLOUD

The main attributes of cloud architectures are:

1. **Multitenancy:** The Centralized Infrastructure is shared between numbers of users. Though the Infrastructure is shared between multiple users in an instance of time, the tenants (users) data is stored in isolation and hence is inaccessible to the other users.

2. **Scalability:** System's ability to increase the capacity of the workload without changing the existing hardware resources. Scalability is possible in cloud architectures as these cloud services depend upon VM to be hosted in the user's machine. These VM are upgraded according to the requirements. Based on workload the VM can be shifted to a different server or are hosted over multiple servers at a given instant of time. Thus, a better system performance can be obtained with existing hardware resources in real. This is static upgradation of system.

3. **Elasticity:** System's ability to accommodate the changing workload dynamically. An elastic system automatically adapts to the change. This is dynamic upgradation of system. This is beneficial when there is a rapid upsurge in the workload or traffic, but eventually the upsurge would stabilize after a point of time. In such a scenario, if too many or too few resources are allocated then either there can be a draft of resources or the resources remaining idle for a long period of time. By this dynamic upgradation we ensure the resource usage is in complete accordance with the load or the demand.

As the computing requirements upsurge, the focus of cloud service providers is to provide services which are comfortable according to requirements and user-friendly. In a user-friendly system, user is less bothered about the technicalities related to the structure and working of the cloud, making it a more secure platform for user's huge amount of data. To increase the quality of the services and infrastructure provided by the cloud, its attributes are expanded for better quality of service to make it more user responsive. But these expansions can lead to compromising on security matters.

8.2   SECURITY IN CLOUD

Cloud computing contributes to an age where data is stored majorly in cloud servers rather than the local systems. There are chunks of data lying in cloud habitats. This data is of great importance, when a data breach or unauthorized access to data can lead to a huge loss to user. Despite cloud being a boon to the majority of storage problems today are prone to the vulnerabilities relating to security, as we come across real incidents where the technology and computing capabilities are misused to intrude cloud or to breach data while it is transported to the cloud. The discussion about security concern is illustrated in the diagram in Figure 8.2 [6].

From Figure 8.2, it is evident that critical issues in the cloud eco-system are related to security incidents. Hence an emphasis is laid to address security-related dynamics of user's data. As we discuss the criti-cality of data security, it is primary responsibility of every CSP to protect clients' data in cloud and there should be some authentication schemes or mechanisms that govern the granting of the access to that data. The client data safety and proper practices and privacy policies are to be validated to assure cloud users of data safety and privacy protection. So, huge invest-ment is obvious for the maintenance of security and privacy in the cloud. The path for the transmission of data through cloud interactions is the internet. The general schemes or protocols like HTTP are unsafe to be used. But even today, there are applications in which the API requests are sent via a HTTP protocol. Hence using such unsafe protocols to transmit data over the internet can pose threat to the integrity and confidentiality of data. In present-day there are intruders and hackers who try to penetrate through the CSP firewall and steal or hamper users' data. This data breach can be very costly to the user and to the CSP. Data either in static (at Rest/ Stored in the cloud) or in transit (Being Transferred to the cloud) is valu-able and of great use. Based on hacking motives, various researchers have proposed variety techniques, to protect security during communication, but the conventional methods fail to ensure complete security. Type of mechanism used for protection depends on the size and nature of data. The diverse datatypes in a cloud ecosystem are [7]:

i.   **Transit Data:** Data in transmission channels between cloud and user.

ii. **Data at Rest:** The static data stored in the cloud.

iii. **Data in Processing:** Data being processed in the server.

iv. **Data Lineage:** Data history which is stored in the warehouses.

v.   **Data Provenance:** History of derivation of data.

vi. **Data Remanence:** The problem of the possibility of recovering the deleted files due to the magnetic properties of the storage medium.

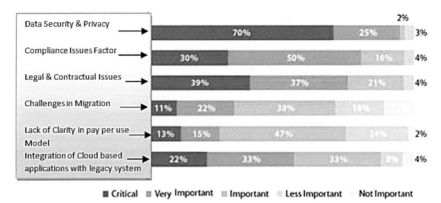

FIGURE 8.2   Critical issues in cloud eco-system – security incidents.

Source: Reprinted/modified with permission from Ref. [6]. © 2015 Elsevier.

In recent times the hacking became a popular profession. National Institute of Standard and Technology (NIST), an agency of United States Department of Commerce, organizes events or competitions in which different security schemes are attacked, based on results the best ones are decided and the loopholes of the other proposed mechanisms (algorithms) are noted. So, when we look at these scenarios, we realize there is an alarming need for security, the necessity that we change the perspective towards technology that ensures security by strengthening it to protect systems from attacks or intruders. There is continuous critical debate for changing of security outlook of cloud technologies and more matured emphasis should be laid on this matter [8]. Even the giant companies who invest huge in making their network foolproof and potentially resistant to the attacks of the hackers, also contract to these cyber-attacks leading to tampering of data, leaking of data or data stealing. To understand cyber-attacks, these are categorized them into major 11 attacks by cloud Security Alliance which are: data breaches, misconfigured system, inability of have potential maintenance of credentials, hijacking, insider threat, insecure api endpoints, structural failures, limited cloud usage visibility, abusive usage of cloud services.

8.3   DATA SECURITY CHALLENGES

For secure cloud, there is a requirement to comprehend data security challenges. A systematic analysis of challenges is indispensable to achieve the objective of making cloud highly secure. Towards this classification of various challenges involved at multiple levels to implement attack resistant security system through proper technological approach is inescapable.

8.3.1 VIRTUALIZATION

Virtualization is the process of creating a virtual version or a virtual platform of a server, a device, or a resource. In simple terms, virtualization is the logical division that can support in usage of resources and services through time sharing. Generally, to facilitate process of the virtualization hypervisors are involved on host machines to run guest operating systems. So, ensure that hypervisor is not vulnerable because the failure in proper functioning, hypervisor can lead to the collapse of entire process and thus endanger data. Another risk involved is if hypervisor fails to maintain proper mechanism to maintain proper allocation and de-allocation of the resources then there is a possibility that after one of Virtual Machines (Virtual Machine-Guest machine) access is ceased and if resources are still available on the system, then is available to next VM running and these prior resources to current VM may be of no importance. But data is obtainable to the current VM which has no need of it, so this instance of data privacy disruption. There can be many security and privacy breaches in data at the virtualization level. And if the hypervisor is compromised if the hacker also gains root permission, then this would lead the hacker having full access to all the resources and operations. This is resolved by having strong authentication before allocation and de-allocation of the resources.

8.3.2 MULTITENANCY

There are shared resources between multiple users simultaneously. So, there exists a possibility that the private data of one user gets leaked and is accessible by any other user. And this condition can be very sensitive to the motives of the hackers. Authenticating users before providing them the access to cloud resources is vital. And another trouble that arises due to multitenancy is: access of data, belonging to a user, shouldn't be granted to another user. Potentially attacks are carried by implementing malicious code over users end, and the access permissions can be reset. Hence, user would be entitled to access other user's resources. Hence it becomes an un-ignorable requirement that we store data in segregation. Generally, data segregation vulnerabilities can be detected by performing testing like Injection attacks (SQL).

8.3.3   STATIVITY, TRANSITIVITY, AND REMINISCE OF DATA

Data saved in the cloud is static, and because it is maintained on a server (cloud) for an extended period of time, maintaining data accuracy is critical. Transit data is information that travels from and to the cloud. Usernames and passwords are the most common types of transit data, and they are always sent to the cloud via APIs. As a result, data must be safeguarded while going via a network for authentication, and this is done utilizing the various encryption techniques. The main encryption techniques pass the credentials via authentication schemes like Session-based authentication or the token-based authentications. The backbone technology and algorithms that govern mentioned schemes are the Cryptographic algorithms. Data reminisce relates to the problems of storages in which even if file has been deleted from file system and the trash permanently, it can still be present or are fetched again by recovering. This problem exists as magnetic properties of storage mediums support retrieval. So, capability to access deleted files is very dangerous to the security of the system. Consider older logs or files, which contain important information, were deleted but these files are accessible and entire information about previous logging and transactions is now open and vulnerable to exposure.

8.3.4   LOCATION TRANSPARENCY

Rather than physical location, the resources are identified based on network configuration. However, this poses a security risk because intruders can alter network settings.

8.3.5   INTEROPERABILITY

Different cloud systems collaborate and connect to work together, which is referred to as interoperability. Data services of collaborating cloud systems are also accessible to interconnected systems. They also have a firm grasp on other systems' authentication procedures, formats, and so on, and as the need for interconnected clouds grows, there is an alarming need to respond to these security issues of interoperability in order to assure safe inter-cloud systems [10]. Hence to transcend the aforementioned challenges, we try to secure data communication channels between cloud and user. Additionally, to ensure foolproof security, arises a necessity that data being transmitted over the channels is encrypted. Cryptography, powerful technique, which supports and resolves issue of security by encrypting data. Cryptography encrypts data making it terribly difficult to understand without reversing the encryption, i.e., decrypting data. This nature of cryptography makes data breach hard even to the hacker. Even if hacker gains access to data and steal it, the hacked information is incomprehensible until it is decrypted and converted back to its original format. So, in such conditions, despite data hacking, no information is revealed to hacker. In the next section, let us overview Cryptography relevant techniques and their applications in securing cloud architectures.

8.4 CRYPTOGRAPHY AND CLOUD SECURITY

8.4.1 CRYPTOGRAPHY: OVERVIEW

Cryptography can be understood as the study and practice of the secure communication techniques in a scenario involving adversaries like presence of the third parties. Systems which implement the cryptographic techniques to offer data security services are termed as the "cryptosystems." There are mainly two types of cryptosystems:

i. **Secret-Key Cryptography (SKC)**
    (Symmetric encryption)

ii. **Public-Key Cryptography (PKC)**
     (Asymmetric encryption)

SKC uses a single key for encryption and decryption of data. Contrastingly, PKC uses two types of keys: public-key, private-key. Cloud service providers utilize public key to encrypt user data and this key is accessible to all in public domain, whereas private-key is utilized by users for decrypting the encrypted data back into its real form, this key is offered privately by CSPs to user using cloud service. Secret-key algorithms are as well called symmetric algorithms while Public-key algorithms are as well called asymmetric algorithms.

8.4.2   CRYPTOGRAPHIC STORAGE IN CLOUD: OVERVIEW

In cryptographic storage system, there are three checkpoints in the implementation [7]:

- Data processor;
- Data verifier; and
- Token generator.

The process followed in a cryptographic storage system is: When user tries to upload to cloud, it is first uploaded to data Processor which encrypts data and is then stored in cloud, in encrypted form. For integrity validation during an upload or a retrieval, data Verifier is used. The Token Generator generates the token which is sent via request to access files. If this token is the legitimate one then, permitted to access and retrieve the files.

8.4.3   CRYPTOGRAPHIC AUTHENTICATION IN CLOUD: OVERVIEW

SAML standard to authenticate users is extensively used by cloud providers. The SAML standards is single sign-on technology (SSO) in which user is authenticated once and these authentication credentials are later communicated to other applications hosted in cloud or the server. The working of the SAML involves three parties:

1. **Principal/Subject:** User requesting to access cloud-hosted application.

2. **Identity Provider:** A service whose prime responsibility is to authenticate user, manage the action associating to the authentication and to store user's identity.

3. **Service Provider:** A cloud-based application, which user intends to use. Every hosted application may have its own authentication. So, generally when a user tries to access a

particular service, they are expected to provide credentials to every application. But through the SAML standard once authenticated through SSO, user does not have to re-authenticate them through the authentication systems of every individual service which they desire to use.

These authentication standards are backed up the cryptographic algorithms (mainly AES-256). The rise in awareness and concerns regarding information security in cloud computing, usage of security algorithms in system processes related to security of data has increased manifold. Cryptographic algorithms play a pivotal role in offering security to data in cloud computing. The unauthorized users are incapable of accessing data as these algorithms hide it from them. There are algorithms which play a significant role in the encryption of data. The next section presents brief insight about different aspects of cryptographic encryption algorithms is presented.

8.5 CRYPTOGRAPHIC ALGORITHMS AND THEIR APPLICATIONS IN CLOUD SECURITY

8.5.1 SECRET-KEY CRYPTOGRAPHY (SKC)

At the time of encryption and decryption, just a single key is used in these techniques. The symmetric key is kept by both the CSP and the user. As a result, the main problem here is implementing key dispersal. This cannot assure safe key transfer over an insecure channel, since there are no unique keys for the user or the CSP. As a result, it's difficult to keep track of data transfers. We can only identify the activity; it is impossible to follow or trace who performed what action during the data transaction process. Popular SKC algorithms are reviewed next, as well as their new application to the security arena.

8.5.1.1 DATA ENCRYPTION STANDARDS (DES)

DES stands for data encryption standards. This algorithm is a popular encryption algorithm. It operates on 64-bit sized plain text blocks and returns cipher text blocks of identical size. The key-length is initially 64 bits but as the algorithm processes, every 8th bit is discarded to form 56 bits long key. This has four modes of operations. This is implemented based on Feistel block ciphers [12]. In this method, permutations of the texts are tried and form boxes on which various iterations of encryption is performed. This method is defenseless to many powerful attacks. Zameer et al. [18] proposed a system in which audio steganography was carried out by building a model built on DES algorithm. In current work, audio is first encrypted using DES algorithm, and then from the encrypted text every LSB of each byte is made zero. The entire information of 8 bits is hidden in the LSB. In the encoding stage, firstly size of audio file is encoded, based on computed size length number, the decoder fetches same numbers of corresponding LSB bits from remaining length of data. The gathered LSB bits are decoded and then the message is decrypted for the original message. The result interprets that a secured level of steganography is done of the audio, but the quality is highly dependent on the size of the audio file. This type of steganography could be implemented to store data on a cloud server. Through these types of techniques, we ensure that data is masked by some other data form making it tough to even identify the original data which is valuable. Such masking of information is done based on numerous types of algorithms.

8.5.1.2   ADVANCED ENCRYPTION STANDARD (AES)

Advanced encryption standard (AES) is a symmetric-key encryption standard in cryptography that is also used to secure cloud data. It was founded in 2001 by the National Institute of Standards and Technology (NIST). The block size of each cipher is 128 bits, while the key sizes are 128, 192, and 256 bits, respectively. AES algorithm was discovered as a complement to DES, as its working mechanism was similar but the prior one performed better on security standards relatively to the later one. AES is an algorithm implemented at both hardware and software levels. This algorithm provides broader scope for developers and researchers to contribute the mathematical modifications/deviations in the algorithm making it deployable for service requirements. Vishwanath et al. [11] proposed and implemented a security system built on RSA and AES algorithms as this was a hybrid cryptography-based system its perfor-mance was better than the pure cryptographic systems (based solely either on RSA or AES). Considering a grouping of security algorithms makes the system immune to vulnerabilities.

8.5.1.3 BLOWFISH

Blowfish [1] is a symmetric key block cipher created by Bruce Schneier in 1993. It takes into account a 64-bit block size as well as varied key lengths ranging from 32 to 448 bits. It employs a 16-round festal cipher block with unusually big key-based s-boxes. Smarajit et al. [12] created a solution that uses a weighted hybrid blowfish algorithm to strengthen cloud security. The proposed way to state provides data central authority and protection in semi-trusted Service Provider scenarios.

8.5.2 PUBLIC-KEY CRYPTOGRAPHY (PKC)

This cryptography was primarily created to support two-party communication, i.e., safeguarding communication across an insecure channel by the correct encryption of randomly generated session keys, which are normally encrypted with the public key and decrypted with the private key. Both keys are linked mathematically. However, knowing one key does not make it possible to know the other. The behavior of the mathematical functions employed in PKC makes obtaining the inverse of functions computationally highly difficult and time-consuming. As a result, we may maintain security by employing two distinct but related keys, notwithstanding the computational overhead. Because distinct keys are utilized, it's simple to keep track of cloud transactions because the encryption and decryption operations are done by separate keys.

8.5.2.1 RSA ALGORITHM

Three mathematicians, Ron Rivest, Adi Shamir, and Len Adleman, proposed this technique in 1977. The name RSA is taken from the first letters of the author's names. This is used in systems where symmetric keys or digital signature authentications must be exchanged. The RSA algorithm is asymmetric and public-key. RSA is a three-step block cipher that involves key creation, encryption, and decryption. For RSA, NIST recommends using keys of a length of 2,048 bits. More advantages include the fact that this is a homomorphic encryption technology in terms of arithmetic operations. That means that if any arithmetic operations are performed on cipher text, the result is a cipher text, which when decrypted is equivalent to (result of an operation performed on simple text). The result is a new cipher text that, when decoded, is equivalent to operations performed on plain text [14]. As concerns about cloud security grow, security systems are being constructed around the concept of homomorphic schemes, in which data is decrypted only by the owner of the private key. Because data transfers and access to requested data are delivered in encrypted form, this strategy is appropriate for cloud environments. The danger of the plain text being accessed by an intruder being managed. In recent years, the scope of homomorphic encryption has expanded to include logical operations, as demonstrated, and proposed by Yi-Fan et al. [14]. In this chapter, composite order bilinear groups are used to make the scheme homomorphic in terms of logical operations such as 'AND' and 'OR' operators. Plain text is translated to ASCII values before being turned to a bit string. And now this bit string is treated as the decimal quantity 'M,' which is a very large number in general.

Vishwanath et al. [15] showed implementing a security system founded on RSA. In currently referring work, security system was applied in two modules: Uploading and Downloading. Where a key generation mechanism was designed to generate secret key from RSA and AES. The keys stayed with the user. Whenever user intends to upload or download, authentication is done based on these keys. RSA proves as a competent security algorithm but has a few limitations like its nature being deterministic, i.e., if the prime numbers are known, then other security parameters and keys could be computed easily. The mathematical and the computational difficulty for the Cryptanalysis is introduced in the step where the chosen prime numbers are being computed and checked. Another limitation of this conventional algorithm is that its operations are homomorphic hence generally even if algorithm is implemented multiple times on same plain text, output remains same. A consequence of this behavior is, encryption schemes are vulnerable to Chosen Cipher Attacks. For initiating Chosen Cipher Attacks, cryptologists gather the information involving the decryptions for various plaintext, then based on collected information of distinct sets of plain texts and their decryptions, they compute secret keys. But these limitations are overcome if different encryptions can be generated for plain-text for every encryption over the same data. This is attained by the padding of the plain texts with padding schemes like OAEP, Selvakumar et al. [16] proposed a padding technique called the optimal asymmetric encryption padding (OAEP), used to pad RSA algorithm to improve safety standards of the system. This model concludes that the OAEP resists the attack on cipher texts.

The main application of this algorithm is to encrypt session keys generated at random by systems that are supported by the SKC or hash function-based security algorithms, i.e., for example in simple terms the public-private keys are used to protect process of distribution of symmetric keys if system has a SKC based security system. This algorithm works efficiently because the selected p and q are very big; hence it's very difficult to computationally evaluate public key and private key just out of 'n' as finding prime factors of any big number is computationally very costly. The problem of finding the prime factors for a given large number complexity wise is a hard problem.

8.5.2.2   ELLIPTIC CURVE CRYPTOGRAPHY (ECC)

Elliptic curve cryptography (ECC) is an asymmetric cryptography algorithm. The algorithms proposed before were mainly the RSA based once, i.e., based on integer factorization problem (IFP) problem. But as the technology is advancing, we need stronger algorithms. Then came few algorithms that were based on stronger and highly complex mathematical concepts like the discrete logarithmic problems (DLP) based EIgamal cryptosystems [17] but later the need arises for more stronger system so led to divert focal point of cryptographers to mathematics of the Elliptic curves.

ECC is based on the mathematical problem of elliptic curve discrete logarithm problem (ECDLP) that is more complex to solve than the IFP or DLP. Currently, there is absence of computationally viable methods to find answer to ECDLP-based problems. There are two keys involved: (i) public key, and (ii) private key.

Comparatively to the other types of PKC algorithms proposed until then ECC demonstrated to be additionally efficient over them. It is algorithm consuming variable length based key encryption. And major gain was, ECC consumed shorter bandwidth for transmitting smaller sized encrypted messages. So, the bandwidth is saved assuring the probability of losing data is less [15]. Alowolodu et al. [4] in his work suggested means of using ECC for securing data over cloud platforms. But in cloud environment data integrity of blocks is hard to be confirmed or checked at users end since data is huge and generally application users lack the technical knowledge for the validations. So, to tackle this limitation, a model was proposed which involved third-party verifiers in systems. Syam et al. [18] proposed a study that used ECC and Sobol techniques to verify the correctness of data saved in the cloud. This project provided a safe storage mechanism for checking the data integrity of blocks by selecting them at random from a cloud server. Because most users lack technical skills, the approach includes a third-party validator other than the user and the CSR who checks for data integrity and safekeeping in place of the users. This strategy ensures a user-friendly and secure environment.

Cryptographic algorithms are traditionally the conventional algorithms, but with growing computational abilities these algorithms remain not adequate for ensuring a completely secure system. Hence this leads to a scenario where we understand, the necessity to rely upon newer and enhanced methods whose purpose is to achieve security concretely. To cater to these needs machine Learning algorithms are regarded as another set of solutions in the field of security. Recently, security systems are mainly hybrid systems where machine learning models are stacked alongside the Crypto algorithms. The cryptosystems have a straightforward function of encryption and the reversal of the function applied gives decrypted output. This characteristic of these algorithms makes them discrete in their way of operation, i.e., there is zero randomness in their working. Hence if the function is known, then input and corresponding outputs decoded are consistent. This consistent nature with no element of randomness or zero-probabilistic nature can pose a danger to the security of the system. We aim to make systems difficult to decode by introducing some level of randomness in systems meanwhile performance of the system is not affected. This is accomplished through algorithms of Machine Learning using which we train security systems built on pattern of data rather than only relying on a definite mathematical function.

8.6 MACHINE LEARNING/DEEP LEARNING-BASED APPROACHES

Machine Learning Models are generally developed for data processing. But usage of ML/DL algorithms for safety applications is considered as innovative approach that unites the subsequent domains: Machine

Learning/Deep Learning and Data Security. As the computational capabilities of computers increase, it becomes easy to resolve and find solutions to computationally hard problems, which are the foundation of cryptographic security algorithms. So, it is critical to discover newer methodologies, not dependent on mathematical functional operations but rather concentrating on the pattern in which data is stored rather than creating cipher texts to store rather try to learn the pattern of data to store it in different format and reconstruction of data is possible based on Training of model (learnt knowledge). So, in these kinds of approaches Machine Learning can aid as technology that protects our data. Some novel machine learning-based cloud security systems are conversed below.

Mbarek et al. [19] proposed a model which strengthens cloud security (for scenarios of Image Uploads) by segmenting the image in four different regions based on intensity levels by a hybrid model of SVM and FCM. Initially ,the images are taken and all pixels are classified in different regions by the SVM linear classifier but as the classifier is mainly intended for supervised learning, we association it with FCM that trains the SVM to progress performance of linear classifier. The FCM plays a main role in extracting the pixel level color features. These features are input vectors that are fed to SVM for classification purpose. Figure 8.3 illustrates the overview of the proposed architecture of model proposing the segmentation model for images to be stored over the cloud.

SVM classifies the pixels into different regions. The image is stored overcloud in its segmented format. This work was carried on healthcare industry images. The security here is enhanced because we focus on learning the pattern of data itself. After we learn the pattern of data, we are arranging data in improved (format different from original) manner over cloud. Hence this caters to issues of security and storage of cloud. The performance of machine learning models to contribute in the direction of security depends on the ability to learn patterns of data. Hence the key aspects while designing such systems is that type of data plays a foremost role. Generally, based on factors like nature and size of data, types of machine learning approaches are adopted. There are numerous data formats, but we aim to ensure its security irrespective of its format. Generally, the major formats in which data exists is either as images or plain text sequences. So, depending on the form of data selection approaches differ. It is observed if the format of data is textual sequence form, then the GA are good to learn them. Shalu et al. [20] established a methodology based on evolutionary algorithms to protect and improve cloud security levels. Data is first transformed to ASCII characters, after which it is turned to a block of bits. The data owner performs the conversion of ASCII characters into a block of bits (DO). To encrypt these blocks of bits and generate cipher blocks, we use genetic algorithms at random. To encrypt these blocks of bits, genetic algorithm processes such as cross overs and mutations are used. The pseudorandom number generator function determines the type of operations to be performed. Once these blocks have been encrypted, they are kept in various locations on the cloud, and the security of these cipher blocks is further enhanced by the DO supplied key. As a result, this system only has one key and the rest is dependent on GA. As a result, the performance of this framework is generally excellent. The decryption is based on dynamic links and reverses the GA-based processes, which are executed by transforming back to ASCII text. The GA works with random data chunks, which adds to security. We tried to compare different strategies in the next part based on security strength, integrity, execution speed, and other criteria, but each cryptographic strategy has its own set of strengths and drawbacks. Machine learning methods may be efficient to learn security pattern, but the ability and competence of the system can be increased using more complex techniques or by using complex model architectures inspired by Deep Learning techniques. However, a new technique to deep neural network training is to train them on encrypted data and learn the pattern of encrypted data rather than the original data. The main notion here is homomorphic encryption, which states that any action done on encrypted data produces a cipher text as a result. On decryption, this cipher is similar to plain text generated by doing the same operation on plain text, hence we can't perform sophisticated computations or operations on cipher texts, as mathematically shown [21]. Ehsan et al. [22] created a model based on deep neural networks that were trained on encrypted data and then tested for classification tasks. By approximating the complex activation functions to their polynomial equivalent while maintaining the model's performance accuracy, deep Neural networks were brought into the realm of homomorphic encryption. Crypto Nets fared worse than the suggested system. As a result, the suggested approach addresses the risk of using cloud platforms for analytics jobs. Using the previously indicated methodology, analytics may be done on cloud servers without compromising data privacy because analytics are run on encrypted data. It is not necessary to have access to encrypted data in order to execute analytics.

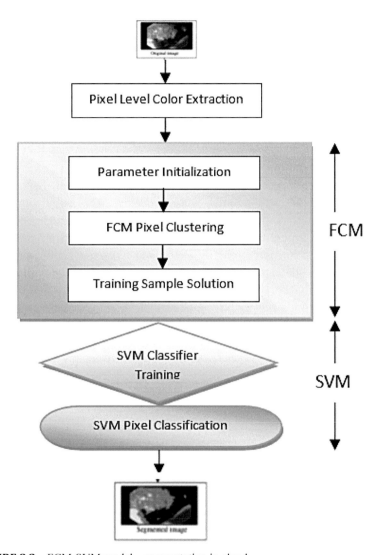

FIGURE 8.3 FCM-SVM model – segmentation in cloud.

Source: Reprinted with permission from Ref. [19]. © 2018. Elsevier.

8.7 ANALYSIS AND COMPARISON

When we compare the various methods stated above, we can claim that asymmetric cryptographic algorithms are a little slower than symmetric cryptographic algorithms, so ECC and RSA are slower than DES and AES, which are both symmetric. This is true because asymmetric algorithms use public and private keys that are larger in size. Similarly, because different keys are used in encryption and decryption, we may say that all asymmetric cryptography approaches are more dependable. Taking storage capacity as a starting point, it is true that algorithms with longer keys require higher storage capacity. However, we can't apply this comparison to all algorithms based on generic claims. As a result, the following is a full comparison among the all methodologists.

The methods listed in Table 8.1 were implemented and their average encryption and decryption times were measured. This simulation simulates the performance of DES, AES, RSA, and Blowfish implementations using built-in Cryptography modules/packages in Python 3 environments. This implementation is thoroughly tested and is optimized for maximum performance for the algorithm. The simulations are conducted with a 64-bit processor with 8 GB of RAM. The development environments used for simulations are Anaconda Jupyter Notebooks and the Google Colab. Considering different sizes of data blocks (20 B to 400 B) the algorithms were evaluated in measure of performance. All the implementations were exact for relatively fair and accurate results. The outcomes recorded in Table 8.1.

TABLE 8.1   Execution Time: Encryption + Decryption Time

Method Proposed	30 Bytes	50 Bytes	100 Bytes	200 Bytes	300 Bytes	400 Bytes
RSA-OAEP (Key: 1,024 bits)	0.003	0.0035	0.0037	X	X	X
RSA-OAEP (Key: 2,048 bits)	0.009	0.015	0.017	0.018	X	X
RSA-OAEP (Key: 3,072 bits)	0.037	0.040	0.044	0.051	0.045	X
DES (using audio steganography)	0.076	0.83	0.84	0.842	0.842	0.834
2-tier (RSA + Blowfish)	0.25	0.267	0.267	0.271	0.317	0.305
3-tier (RSA + Blowfish + DES)	0.303	0.305	0.31	0.343	0.35	0.35
Hybrid system (AES + ECC)	0.0049	0.005	0.0057	0.0057	0.0057	0.0057

Table 8.1 shows the aggregate time for encryption and decryption for the methods proposed. The units of time are seconds. The conclusions depicted through the simulations are: ECC is most amicable when computational power, memories are limited. Moreover, as the size of the key increases in RSA, permutations of cracking the key also increases. Each technique has its individual importance depending upon the computational specifications provided. The overall simulation result interprets that as we increment the size of text, a larger key is desired for encryption. But this is computationally costly as the text size can keep growing invariable; hence the other encryption methods that have a unique approach like the hybrid cryptosystems or other masking methods like video steganography can serve in these situations. To increase security algorithm's performance their complexity is upgraded, consequently increases time of encryption and decryption. Hence here we encounter the necessity for inventing a data security system not based on cryptographic algorithms purely. As the computational power of processors increase probability of finding solution of mathematical functions also increases. We try to understand the nature of algorithms to custom them efficiently.

8.7.1   BENEFITS AND DRAWBACKS OF DES

The behavior of the algorithm is complicated, and it's difficult to decipher the mathematical rationale behind it since the Boxes go through a permutation-combination phase before moving on to the next round of creating cipher letters. In their research, Sombir et al. [23] demonstrated that DES outperforms RSA in terms of data encryption and decryption time.

Due to the permutation combination phase, it's possible that a given pattern of two inputs will produce the same output. As a result of the presence of the permutation and combination phases, there is a disparity between the system's initial and end outputs.

8.7.2   BENEFITS AND DRAWBACKS OF RSA

It entails calculating the integer factorization [24] of a larger number that is difficult to compute. Finding the integer factor is a time-consuming and laborious task. As a result, the size of keys should be larger in order to be more secure. Even if one of the keys is known, the other cannot be deciphered since they have a mathematical relationship.

As computer processing power grows, so does the amount of data that can be processed. And concepts like parallel computing result in an exponential increase in calculation abilities, rendering RSA vulnerable to these breaches because intruders use the brute force method (to generate all the possible set of keys and test each key). To improve security, larger numbers are expected to be considered as a choice for the number whose factorization is conducted, i.e., larger keys are generated to overcome increasing computing powers, but maintaining larger keys is problematic because decryption is a lengthy procedure.

8.7.3   BENEFITS AND DRAWBACKS OF AES

At both the hardware and software levels, it is simple to implement [25]. This standard is used by the majority of open-source solutions to protect data. To break security standards with 256 bits, around 2,256 attempts are required. Brute-force key breaking is computationally expensive and time-consuming. As a result, it might be considered a safe protocol.

It makes it vulnerable to key scheduled attacks because it uses too simplistic mathematical logic, such as simple algebraic structures. As a result, we find that the algorithm's straightforward mathematical formulation is weak. As a result, it's simple to break into and infiltrate a system secured by this algorithm.

8.7.4   BENEFITS AND DRAWBACKS OF ECC

It runs extremely quickly because binary curves are well-suited to hardware [26]. The significance of this technique is that it maintains security with smaller keys and that the encryption size is not fixed, i.e., smaller messages result in smaller encrypted communications. The size of the message is proportional to the encryption size. As a result, this algorithm is adept at efficiently utilizing bandwidth.

Standard curves, in particular, make securing ECC difficult. When compared to Schnorr signatures, ECC Standards are mostly imitated. The key is tampered with since the signature is done with a faulty random number generator. This theory has some flaws, particularly when it comes to binary curves.

8.7.5 BENEFITS AND DRAWBACKS OF BLOWFISH

DES is slower than this. On small memory spaces, the technique works well [27]. The ability to change the size of the keys [28]. By increasing the number of rounds, you can develop your strength.

Because each user has their own key, managing keys becomes more difficult as the number of users grows. The decryption method has a disadvantage over other algorithms in terms of execution time and serial throughput.

8.8 THREAT TO CLOUD SECURITY: QUANTUM PERCEPTION

Though Cryptography or machine learning may cater to the need of having secure communication, these do not ensure a permanent solution to cyber threats. With advancing technologies, the emergence of quantum computers would mean increasing the computational ability but simultaneously risking our encryption schemes. Most of encryption algorithms are adhered to by the complex mathematics which in practicality results in computationally costly process. But with quantum computers, these mathematical functions which are computationally challenging now, might seem solvable and programmable without being computationally costly. Hence it means easy to crack keys of existing encryptions. Currently, quantum computers are not fully functional but possibilities of such technology emerging sooner into actual world is high. So, plans for implementing security schemes resistant to quantum powers are desired. To secure traditional cryptography from quantum powers there are two propositions:

- Quantum cryptography; and
- Post-quantum cryptography.

8.8.1 QUANTUM CRYPTOGRAPHY

Quantum cryptography relies on different principles of quantum physics rather than only on mathematical formulations to ensure security. This cryptography is theoretically impenetrable. In quantum cryptography, key is shared via entangled-based protocol meaning key is transferred by quantum states which are in superposition state. This cryptography is stronger transcending traditional cryptography as it implements key transfer by being certain about third parties. If an attacker tries to intervene then quantum states are no more in superposition state meaning the attacker cannot copy information or operate on copied state.

The main application of quantum key transfer is communication over highly stretched distances using quantum properties. The popular protocols of quantum key exchange are discussed in subsections.

8.8.1.1 BB84 PROTOCOL

In BB84, the secret key is encoded using the polarized states of the photon. This feature of encoding using photon energy makes the system secured as photon properties safeguard keys from attack. If polarization state of photon is measured then its consequences in destroying of photon completely making the attack difficult for eavesdropper. To penetrate system, eavesdropper is forced to send new photon, by default in wrong polarization state, leading to reveal of attack.

> **BB84 Application to Cloud Security:** Yasser et al. proposed system which encrypts user data by using Enhancement BB84 protocol generated key. Data is decodable by authenticated user who provides key over quantum channels. This method implementation resulted in system ensuring security for confidentiality of data. Observations reveal less encryption time increasing performance and data transfer rates in cloud systems. This betterment of time period and performance is observed as the quantum protocols are secured and faster over the traditional Crypto methods.

8.8.1.2 E91 PROTOCOL

This protocol uses entangled state of photons to transmit information between two parties. Protocol mainly follows fundamental concept, the quantum states distributed between parties are correlated, meaning they are directionally synchronous, i.e., either vertically polarized or horizontally polarized. Attacker trying to intrude into this set up leads to destruction of correlation present between these quantum entangled states leading to reveal of attack to parties involved in communication.

➢ **E91 Application to Cloud Security:** This protocol is appropriate for cloud environments as they ensure high security. Li et al. [29] presented analysis about this protocol and observations revealed that when an eve (attacker) attempts to interfere, this interference can be detected by entitled parties involved in communication. Although Eavesdropper drops eve, he/she could access only up to 50% of raw key data and is unaware about which part of the key does this 50% belongs to. For optimized performance of protocol is observed when noise levels are relatively low.

From the perspective of hardware, for implementation of quantum cryptography, there are multiple challenges which make application of quantum cryptography difficult. To resolve hardware level limitations, the need arises to discover better techniques like Post-Quantum Cryptography which are practically implementable.

8.8.2   POST-QUANTUM CRYPTOGRAPHY

The aim of quantum cryptography is to strengthen traditional cryptography schemes rather replacing them. In quantum cryptography focal point is key transfer. Implementing this cryptography is relatively viable at hardware level compared to quantum cryptography as there is zero requirement of specialized quantum hardware. The fundamental basis this cryptography is mathematically hard problems, which are immune to identified quantum attacks, are utilized to encrypt data. There are three approaches proposed:

- Lattice-based cryptography;
- Error correcting cryptography; and
- Multivariate polynomial-based cryptography.

Among the three afore-mentioned approaches, Lattice-Based Cryptography is maximum stable. Cloud ecosystems are expected to have agility and low latency, incorporation of these features through the traditional cryptosystems is impossible hence, Nejatollah et al. [30] proposed an accelerated system, proficient enough to run multiple cryptographic algorithms. This system was inspired from application-specific integration circuits (ASCI).

8.9   CONCLUSION

In this chapter, numerous cryptographic methods for encrypting data were reviewed, as well as various types of ciphers that can increase cloud security, regardless of whether data is static or in transit. As new technologies emerge, newer ways to cloud security are embraced, as threats and breaches become impervious to previous tactics. As a result, cutting-edge machine learning-based implementations for cloud data security are also explored. A comparison is made between the algorithms, and the best technique for implementation is chosen depending on the needs of cloud systems. Currently, modern era is looking forward to the emergence of quantum computers and super computers. Reflecting on security in accordance to their emergence posses' threat to existing systems of security. To surpass this threat, strengthened version of traditional cryptography called quantum cryptography and post-quantum cryptography were proposed. In this survey, a brief understanding about quantum cryptography and quantum cryptography with supporting application to the field of cloud security. In this era of the data revolution, there is a great amount of data to manage and store, which draws consumers' attention to cloud servers. As a result, maintaining the secrecy and security of cloud data is a critical part of ensuring the long-term viability of prospective cloud services.

KEYWORDS

- **cloud computing**
- **cryptography**
- **deep learning**
- **machine learning**
- **post-quantum cryptography**
- **quantum cryptography**

REFERENCES

1. Schneier, B., et al., (1994). Description of a new variable-length key, 64-bit block cipher (Blowfish). *Fast Software Encryption, Cambridge Security Workshop Proceedings* (December 1993) (Vol. 19, No. 4). Springer-Verlag, 1994. This paper also appeared as: The Blowfish Encryption Algorithm. Dr. Dobb's Journal.

2. Griffin, R., (2018). *Internet Governance.* Scientific e-Resources.
3. Corbató, F. J., (2012). An experimental time-sharing system. *SJCC Proceedings.* MIT.
4. Alowolodu, O. D., Alese, B. K., Adetunmbi, A. O., Adewale, O. S., & Ogundele, O. S., (2013). Elliptic curve cryptography for securing cloud computing applications. *International Journal of Computer Applications.*
5. Rich, M., (2009). What's in a name? *Utility vs. Cloud vs. Grid.* Retrieved from: Data Center Knowledge: Chapter 7, Cloud Architecture and Datacenter Design (57 pages) in Distributed Computing: Clusters, Grids and Clouds, All rights reserved by Kai Hwang, Geoffrey Fox, and Jack Dongarra, May 2, 2010.
6. Rao, R. V., Selvamani, K., (2015). Data security challenges and its solutions in cloud computing. *Procedia Computer Science., 48*, 204–209.
7. Kelsey, R., (2013). Cloud cryptography. *International Journal of Pure and Applied Mathematics, 85*, 1–11.
8. Vijay, V., (2011). Rethinking cyber security. In: *Proceedings of the 4th International Conference on Security of Information and Networks (SIN '11)* (pp. 3, 4). Association for Computing Machinery, New York, NY, USA. doi: https://doi.org/10.1145/2070425.2070428.
9. Aws, N. J., & Mohamad, F. B. Z., (2013). Use of cryptography in cloud computing. *IEEE International Conference on Control System, Computing and Engineering.*
10. Kiranbir, K., Sandeep, S., & Karanjeet, S. K., (2017). Interoperability and portability approaches in inter-connected clouds: A review. *ACM Comput. Surv., 50*(4), 40. Article 49.
11. Naresh, V., & Thirumala, R. B., (2016). A study on data storage security issues in cloud computing. *Procedia Computer Science., 92*, 128–135.
12. Smarajit, G., & Vinod, K., (2018). Blowfish hybridized weighted attribute-based encryption for secure and efficient data collaboration in cloud computing. *Appl. Sci., 8*, 1119. doi: 10.3390/app8071119.
13. Abbas, A., Hidayet, A., Selcuk, U. A., & Mauro, C., (2018). A survey on homomorphic encryption schemes: Theory and implementation. *ACM Comput. Surv. 51*, 4, Article 79.
14. Yi-Fan, T., Chun-I, F., Ting-Chuan, K., Jheng-Jia, H., & Hsin-Nan, K., (2017). Homomorphic encryption supporting logical operations. *Proceedings of the 2017 International Conference on Telecommunications and Communication Engineering (ICTCE).*
15. Vishwanath, M., & Aniket, S., (2014). *Enhancing the Data Security in Cloud by Implementing Hybrid (Rsa & Aes) Encryption Algorithm.* INPAC.
16. Selvakumar, S., Sahil, B., & Vidit, K. S., (2017). Secure sharing of data in private cloud by RSA – OAEP algorithm. *International Journal of Pure and Applied Mathematics, 115*(6), 689–695.
17. Jia, C., Yudong, Q., Bei, H., & Qinghua, C., (2016). Research on cloud computing data security based on ECDH and ECC. *International Journal of Simulation: Systems, Science and Technology.*
18. Syam, K. P., & Subramanian, R., (2011). An Efficient and Secure Protocol for Ensuring Data Storage Security in Cloud Computing. ISSN (PRINT): 2393-8374, (ONLINE): 2394-0697, *5*(3), 2018.

19. Marwan, M., Kartit, A., Ouahmane, H., (2018). Security enhancement in healthcare cloud using machine learning. *Procedia Computer Science.*

20. Shalu, M., & Sushil, K. S., (2018). A new security framework for cloud data. *Procedia Computer Science.*

21. Pengtao, X., Misha, B., Tom, F., Gilad-Bachrach, R., Kristin, E. L., & Michael, N., (2014). *Crypto-Nets: Neural Networks Over Encrypted Data.* CoRR abs/1412.6181.

22. Ehsan, H., Hassan, T., Mehdi, G., & Catherine, J., (2017). Privacy-preserving machine learning in cloud. In: *Proceedings of the 2017 on Cloud Computing Security Workshop (CCSW '17).* Association for Computing Machinery.

23. Sombir, S., Sunil, K. M., & Sudesh, K., (2013). A performance analysis of DES and RSA cryptography. *International Journal of Emerging Trends & Technology in Computer Science (IJETTCS).*

24. Na, Q., Wei, W., Jing, Z., Wei, W., Jinwei, Z., Junhuai, L., Peiyi, S., et al., (2013). Analysis and research of the RSA algorithm. *Information Technology Journal, 12,* 1818–1824.

25. https://www.rfwireless-world.com/Terminology/Advantages-and-disadvantages-of-AES.html

26. https://steemit.com/cryptography/@shubhamupadhyay/elliptic-curve-cryptography-or-rsa-algorithm-and-why-or-advantages-and-disadvantages.

27. https://brainly.in/question/1744703.

28. Sudha, S. S., & Divya, S., (2013). Cryptography in image using blowfish algorithm. *International Journal of Science and Research (IJSR).*

29. Leilei, L., Hengji, L., Chaoyang, L., Xiubo, C., Yan, C., Yuguang, Y., & Jian, L., (2018). The security analysis of E91 protocol in collective-rotation noise channel. *International Journal of Distributed Sensor Networks, 14*(5).

30. Nejatollahi, H., Dutt, N., & Cammarota, R., (2017). Trends, challenges and needs for lattice-based cryptography implementations: Special session. *Proceedings of the Twelfth IEEE/ACM/IFIP International Conference on Hardware/Software Codesign and System Synthesis Companion,* 1–3.

31. Vishwanath, M., & Aniket, S., (2014). *Enhancing the Data Security in Cloud by Implementing Hybrid (Rsa&Aes) Encryption Algorithm.* INPAC.

32. Ayan, M., (2005). *Diffie-Hellman Key Exchange Protocol, Its Generalization and Nilpotent Groups.*

33. Jaspreet, K. G., (2015). *ElGamal: Public-Key Cryptosystem.* A paper presented for Masters Degree, Indiana State University.

34. Zameer, F., & Tarun, K., (2011). *Audio Steganography Using DES Algorithm.* INDIACom-BVICAM.

CHAPTER 9

Role of Hospitality Institutions in Development of Employability Expertise: Students' Perceptions Based on Data Analysis

SANTANU DASGUPTA

School of Hospitality and Tourism, Sister Nivedita University, Kolkata, West Bengal, India

ABSTRACT

The recruitment in the hospitality industry is one of the highest in the globe. Time has changed and the 'soft expertise' are overshadowing 'hard expertise.' During the interview movement, employability expertise outnumber technical expertise. Interviewer are relying more on Psychometric Exam to find the "ideal" applicant with multi-dimensional expertise. At this point, the roles played by universities and institutions are crucial. If a student is found to be "not employable," they face the heat. This chapter aims to compare the perceptual difference the male and female undergraduate hospitality students on employability expertise taught at the institutions. A total of 205 students from eight West Bengal hospitality schools competed in the study. The information was gathered through an online questionnaire-answer mode distributed through a social media site. The data was analyzed using SPSS version 23. According to the findings, there is no substantial difference between male and female

System Design Using the Internet of Things with Deep Learning Applications.
Arpan Deyasi, Angsuman Sarkar, and Soumen Santra (Eds)
© 2024 Apple Academic Press, Inc. Co-published with CRC Press (Taylor & Francis)

hospitality students' perceptions of the industry's employability abilities and the level of feed they accept from their universities.

9.1 INTRODUCTION

The Hospitality Corporation or industry is among foremost global manpower recruiting industries. Time moves at a breakneck speed, and 'soft expertises' are increasingly displacing 'hard talents.' During the job-hiring process, employability expertise outnumber technical expertise. Interviewers are relying more on Psychometric Exams to find the "ideal" applicant with multi-modal expertise. At this point, the activities played by universities and colleges are crucial. Various studies looked at the perspectives of various stakeholders on the nature of employability, including "companies" [1, 2]; "students" [1]; and "workers" [3]. In the 1980s, prominent research Fellows from the United Kingdom, the European countries nearly 1990s, and Australia around 2000 developed numerous employability systems. In those approaches, both talents of technical and/or human characteristics were carried the same weightage.

Employability was described by Hillage and Pollard [4] like "an aptitude to be employable, i.e.: (i) a capability to obtain first employment; (ii) capacity to keep job; and (iii) capability to earn new job if required." Researchers have questioned the curriculum's worth and value [5–9]. According to reports, the researchers have protested vehemently about the discrepancy they discovered between the institution's course content and the industry's requirements.

9.2 REVIEW OF LITERATURE

Wikramasinghe and Perera [10] conducted their research on computer science engineering entry-level graduate jobs in Sri Lanka. They reported that there is an influence of gender of graduates in employability expertise. Employability of an individual depends upon knowledge, expertise, and attitude; as used and deployed by an individual. The researchers stated about the difficulty in finding empirical studies that investigated and compared employability expertise of graduates, lecturers, and employers that they think to be valuable while applying for entry-level jobs. The researchers identified two aspects of employability expertise as: (i) subject expertise;

and (ii) transferable expertise. The changes that have been witnessed in the employment pattern were influenced by employability expertise as transferable expertise. As advocated by the researchers, the primary idea of HEI (higher education institutions) is to impart knowledge, expertise, attitudes, and abilities to the students to empower them with lifelong employability. The employers, students, and lecturers have identified soft expertise like problem solving, self-confidence and teamwork as important features. The graduates identified creative and innovative thinking as an important skill. The lecturers marked oral communication to be important. Universities using industry placement as a mode to address the employability expertise was also seen. As suggested, a graduate who has necessary expertise becomes motivated and efficient in the retention of employment by fulfilling their job task.

Song et al. [11] conducted a study of entry-level individuals with and without disabilities. In the study, the most important five expertise by the professional graduates were Work Ethics, Team Work, Oral Communication, Social Responsibility, and Reading Comprehension. The study aimed to identify the employer's expectation on the topic of research. A survey was conducted among 188 individuals where 168 participated in the study. A four-point Likert Scale was used to analyze the data. The purpose of this analysis was to do the comparison of average rating scores in each group. The highest level of skill seen in employees with disability was displaying of personal integrity/honesty in work, whereas the employees without disabilities rated "ability to read with understanding" as the most important skill. Amongst some noticeable differences in the employers "safety of employees" was given a higher rating for employees with disabilities than to those employees without disabilities. The female employers showed higher expectation than their male counterparts on their employees. They looked for better social expertise, work expertise and personal traits. However, the reason was unclear.

Grashiela et al. [12] worked on the employer's feedback on job performance of the graduates through a descriptive type of research method. They focused on employability of computer engineering graduates in non-government higher education institution in the Philippines. The study aimed to determine the employability of computer engineering graduates in the Philippines by determining the employment status, nature of employment, competencies learned in college and work-related values. The most useful soft expertise as perceived by the graduates were:

- Information technology expertise;"
- "Communication expertise;"
- "Problem solving expertise;"
- "Critical thinking expertise;"
- "Human relation expertise."

Industry partners have shown high regards for the graduates in terms of competencies like skill in research, work discipline, communication skill and computer skill. The students perceived entrepreneurial skill to be of least importance for the first employment similarly the employers found the same skill to be least present among the students.

Oliveira and Guimarães [13] described the competency-based approach for curriculum development to improve the employability of the graduates as an outcome of Bologna Reform process at one university in Portugal. The universities in Portugal started encouraging the skill development of the students focusing on the employability from mid-1990's like other universities in the European Union. A five-stage implementation strategy was designed for competency-based approach towards rearrangement of curriculum:

- ➢ **Stage 1: Market Value Expertise:** The key expertise that is to be developed in the students so as to make them market ready through thorough need assessment of the employers.

- ➢ **Stage 2: Curricular Deconstruction:** Focusing on the learning outcome the curriculum was discussed.

- ➢ **Stage 3: Curriculum Redesign:** Structuring different subjects throughout three years addressing the transferrable expertise was done.

- ➢ **Stage 4: Mapping Transferrable Expertises:** The transferrable expertise were identified in each subject, group of subjects and academic year.

- ➢ **Stage 5: Students' Individual Coaching:** The students were given special coaching on transferrable expertise where further development was necessary.

Dimalibot et al. [14] expressed that the greater number of employees and entry-level graduates should possess knowledge and expertise in their respective fields to remain in competition. In research focusing on the employment status of tourism graduates in the Philippines, the researchers proposed an action plan for the enhancement of the curriculum and services offered at a university. The career planning in the area of tourism gives excellent rewards as this industry is the fastest growing industry in the world, creating a wide range of career options. The employment opportunities in the field require knowledge and expertise related to the field of specialization. As found, the expertise which are learned most in the college are communication expertise followed by human relation skill and critical thinking expertise. The work-related values which are valued most by the employers are professional integrity followed by perseverance and hard work and punctuality.

Boyadjieva and Trichkova [15] aimed to study the relationship between the profile of HEI's and graduate employability in Bulgaria. The economic globalization and knowledge economy has significantly influenced the perspective of graduates' employability. The researchers found that the graduates passing out from private HEI's are no way less employable than the graduates of public HEI's but a notable difference was witnessed about the employability of graduates of different HEI's studying same course. The researchers stated that the employability of the students is highly impacted by their institutional characteristics. The influence of quality academic staff has a huge impact on the employability measures as indicated in the result. The thinking of a student regarding the labor market, competition for jobs and employability is greatly controlled by the characteristics of the HEI's.

9.3   RESEARCH METHODOLOGY

9.3.1   OBJECTIVES

The goal of this study was to see if there were any differences in perceptions of the relevance of employability expertise among male and female undergraduate hospitality students, as well as the "amount of input on employability expertise received at their respective colleges."

9.3.2   DESIGN OF RESEARCH

The Fellow or researcher adapted a quantitative data analysis method. The participants were selected from West Bengal's undergraduate hotel management program. Students from eight West Bengal hospitality schools were involved in the study. Participants in the study were in their final year of hospitality institutions courses.

9.3.3   SURVEY TOOL

The questionnaire that was used for the collection of data was having two parts. The demographic background of the students was conceding the first part of the questionnaire. The second section of the survey focused on the employability expertise taught in undergraduate hospitality courses at the institutions. There were 34 employability expertises in the questionnaire. They were then separated into five skill sets: (i) basic academic expertises, (ii) interpersonal expertises, (iii) technical expertises (personality), (iv) work-related expertises, and (v) social expertises. Traits value of employable abilities was rated using a 5-point Likert Scale with exceedingly non-important (a), non-important (b), normal (c), important (d), and exceedingly important (e).

9.3.4   VALIDITY AND RELIABILITY

The researcher has used an instrument that has been adapted from the research article [16] that was published in IJRAR, Volume 6, Issue 1, pp. 161–166.

9.3.5   HYPOTHESES

Using the objectives and the research questionnaire as a guide, the researcher presented the following null hypotheses:

  ➤ **Hypothesis 1:** When it comes to the institution's input on Basic Academic Expertises, there is no discernible difference between male and female undergraduate hospitality students.

> ➤ **Hypothesis 2:** When it comes to the institution's feedback on Personality Traits, there is no discernible variation in perception between male and female undergraduate hospitality students.

> ➤ **Hypothesis 3:** When it comes to the institution's contribution on Work Related Expertises, there is no discernible difference between male and female undergraduate hospitality students.

> ➤ **Hypothesis 4:** When it comes to the institution's input on Inter-personal Expertises, there is no discernible variation in perception between male and female undergraduate hospitality students.

> ➤ **Hypothesis 5:** There is no variation in perception between male and female undergraduate hospitality students when it comes to the institution's Social Expertises contribution.

> ➤ **Hypothesis 6:** In terms of the relevance of Basic Academic Expertises, there is no discernible difference between male and female undergraduate hospitality students.

> ➤ **Hypothesis 7:** When it comes to the value of personality traits, there is no variation in perception between male and female undergraduate hospitality students.

> ➤ **Hypothesis 8:** In terms of Work-Related Expertises, there is no discernible difference between male and female undergraduate hospitality students.

> ➤ **Hypothesis 9:** When it comes to the importance of interpersonal expertise, there is no variation in perception between male and female undergraduate hospitality students.

> ➤ **Hypothesis 10:** When it comes to the importance of social expertise, there is no variation in perception between male and female undergraduate hospitality students.

9.3.6   DATA COLLECTION

The study was conducted using purposive random sampling. The outgoing batch of undergraduate hospitality students took part in the study. The data was composed via an internet survey instrument. Questionnaires were sent to 235 students and 205 valid questionnaires were received. Out of 205 respondents, 115 were male students and 90 were female. The researcher collected the data between February-April 2021 in West Bengal.

9.3.7   STATISTICAL METHODOLOGIES USED

The independent sample t-test was used to test hypotheses 1 through 10. The Perceptions of male and female students of hospitality institutions are relevant in terms of employability expertise, as well as the level of feed from institutions on such abilities, were compared. For statistical analysis, IBM SPSS Version 23 was employed.

9.4   DATA ANALYSIS AND INTERPRETATION

Statistics reports of the students is presented in Table 9.1.

TABLE 9.1   Ratio of Students Based on Gender

Students	Number	Ratio (%)
Male	127	62
Female	78	38
Total	205	

Source: Primary data.

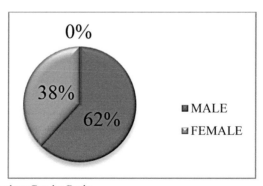

CHART 9.1   Student Gender Ratio

Source: Primary data.

Out of the 205 respondents 127 (62%) were male while 78 (38%) were female. All the students (100%) who have participated in the survey were learning in the last year of their academic process.

➢ **Hypothesis 1: H:** *There is no perceptual difference in the male and female undergraduate hospitality students regarding input by the institution on Basic Academic Expertise.*

The perceptual differences of male and female undergraduate hospitality students (N = 205) about amount of teaching input from the institution for basic academic expertise were compared using an independent–samples t-test, with the significant difference in mean of sampling data at the 0.05 level.

Based on the above result, no "statistically-significant difference" in the scores of Table 9.2 (basic academic expertise) about the perception of male students (M = 3.9, SD = 0.55) and female students (M = 3.8, SD = 0.65) was found. Null hypothesis is accepted.

TABLE 9.2   Independent Samples t-Test Result for Input by the Institution on Basic Academic Expertise

	Group	N	Mean	Standard Deviation	t	df	Sig. (2-Tailed)
Basic Academic Expertise	Male	127	3.9	0.55	1.36	203	0.17
	Female	78	3.8	0.65	1.42	182.91	0.16

Source: Primary data.

➢ **Hypothesis 2: H:** *There is no perceptual difference in the male and female undergraduate hospitality students regarding input by the institution on Personality Traits.*

The perceptual differences of male and female undergraduate hospitality students (N = 205) about amount of teaching input from the institution related personality traits were compared using an independent–samples t-test, with the significant difference in mean of sampling data at the 0.05 level.

Based on the above result, no "statistically-significant difference" in the scores of Table 9.3 (personality traits) about the perception of male students (M = 4.3, SD = 0.57) and female students (M = 4.2, SD = 0.55) was found. Null hypothesis is accepted.

TABLE 9.3   Independent Samples t-Test Result for Input by the Institution on Personality Traits

	Group	N	Mean	Standard Deviation	t	df	Sig. (2-Tailed)
Personality Traits	Male	127	4.3	0.57	1.32	203	0.19
	Female	78	4.2	0.55	1.30	156.90	0.19

Source: Primary data.

> ➤ **Hypothesis 3: H:** *There is no perceptual difference in male and female undergraduate hospitality students regarding input by the institution on Work Related Expertise.*

An independent–samples t-test was conducted to compare the perceptual difference of male and female undergraduate hospitality students (N = 205) concerning the measure of teaching feed from the college regarding work-related expertise with significant difference in mean of sampling data at 0.05 level.

Based on the above result, no "statistically-significant difference" in the scores of Table 9.4 (work-related expertise) about the perception of found. Null hypothesis is accepted.

TABLE 9.4   Independent Samples t-Test Result for Input by the Institution on Work-Related Expertise

	Group	N	Mean	Standard Deviation	t	df	Sig. (2-Tailed)
Work-Related Expertise	Male	127	4.1	0.65	1.69	203	0.09
	Female	78	3.9	0.69	1.72	171.48	0.09

Source: Primary data.

> ➤ **Hypothesis 4: H:** *There is no perceptual difference in the male and female undergraduate hospitality students regarding input by the institution on Interpersonal Expertise.*

The perceptual differences of male and female undergraduate hospitality students (N = 205) about the amount of teaching input from the institution related interpersonal expertise were compared using an independent–samples t-test, with the significant difference in mean of sampling data at the 0.05 level.

Based on the above result, no "statistically-significant difference" in the scores of Table 9.5 (interpersonal expertise) about the perception of male students (M = 4.0, SD = 0.61) and female students (M = 3.9, SD = 0.62) was found. Null hypothesis is accepted.

TABLE 9.5 Independent Samples t-Test Result for Input by the Institution on Interpersonal Expertise

	Group	N	Mean	Standard Deviation	t	df	Sig. (2-Tailed)
Interpersonal	Male	127	4.0	0.61	1.19	203	0.23
Expertise	Female	78	3.9	0.62	1.20	165.24	0.23

Source: Primary data.

➢ **Hypothesis 5: H:** *There is no perceptual difference in male and female undergraduate hospitality students regarding input by the institution on Social Expertise.*

An independent–samples t-test was conducted to compare the perceptual difference of the male and female undergraduate hospitality students (N = 205) concerning the measure of teaching feed from the college regarding social expertise with significant difference in mean of sampling data at 0.05 level.

Based on the above result, no "statistically-significant difference" in the scores of Table 9.6 (interpersonal expertise) about the perception of male students (M = 4.0, SD = 0.61) and female students (M = 3.9, SD = 0.62) was found. Null hypothesis is accepted.

TABLE 9.6 Independent Samples t-Test Result for Input by the Institution on Social Expertise

	Group	N	Mean	Standard Deviation	t	df	Sig. (2-Tailed)
Social	Male	127	4.2	0.70	1.60	203	0.11
Expertise	Female	78	4.0	0.76	1.64	173.72	0.10

Source: Primary data.

➢ **Hypothesis 6: H_0:** *There is no perceptual difference in the male and female undergraduate hospitality students regarding the importance of Basic Academic Expertise.*

An independent–samples t-test was conducted to compare the perceptual difference of the male and female undergraduate hospitality students (N = 205) concerning the measure of teaching feed from the college regarding basic academic expertise with significant difference in mean of sampling data at 0.05 level.

Based on the above result, no "statistically-significant difference" in the scores of Table 9.7 (basic academic expertise) about the perception of male students (M = 4.2, SD = 0.49) and female students (M = 4.1, SD = 0.47) was found. Null hypothesis is accepted.

TABLE 9.7   Importance of Result of Independent Sample t of Basic Academic Expertise

	Group	N	Mean	Standard Deviation	t	df	Sig. (2-Tailed)
Basic Academic Expertise	Male	127	4.2	0.49	1.38	203	0.17
	Female	78	4.1	0.47	1.37	158.66	0.17

Source: Primary data.

> ➤ **Hypothesis 7: H_0:** *There is no perceptual difference in male and female undergraduate hospitality students regarding the importance of Personality Traits.*

An independent–samples t-test was conducted to compare the perceptual difference of male and female undergraduate hospitality students (N = 205) concerning the measure of teaching feed from the college regarding personality traits with significant difference in mean of sampling data at 0.05 level.

Based on the above result, "statistically-significant difference" in the scores of Table 9.8 (personality traits) about the perception of male students (M = 4.6, SD = 0.34) and female students (M = 4.5, SD = 0.48) was found. Null hypothesis is rejected.

TABLE 9.8   Importance of Result of Independent Sample t of Personality Traits

	Group	N	Mean	Standard Deviation	t	df	Sig. (2-Tailed)
Personality Traits	Male	127	4.6	0.34	2.15	203	0.03
	Female	78	4.5	0.48	2.32	197.78	0.02

Source: Primary data.

> ➤ **Hypothesis 8: H_0:** *There is no perceptual difference in male and female undergraduate hospitality students regarding the importance of Work-Related Expertise.*

An independent–samples t-test was conducted to compare the perceptual difference of male and female undergraduate hospitality students (N = 205) concerning the measure of teaching feed from the college regarding work-related expertise with significant difference in mean of sampling data at 0.05 level.

Based on the above result, "statistically-significant difference" in the scores of Table 9.9 (work-related expertise) about the perception of male students (M = 4.6, SD = 0.39) and female students (M = 4.4, SD = 0.57) was found. Null hypothesis is rejected.

TABLE 9.9   Importance of Result of Independent Sample t of Work-Related Expertise

	Group	N	Mean	Standard Deviation	t	df	Sig. (2-Tailed)
Work-Related	Male	127	4.6	0.39	2.50	203	0.01
Expertise	Female	78	4.4	0.57	2.73	200.39	0.01

Source: Primary data.

> ➤ **Hypothesis 9: H_0:** *There is no perceptual difference in the male and female undergraduate hospitality students regarding the importance of Interpersonal Expertise.*

An independent–samples t-test was conducted to compare the perceptual difference of the male and female undergraduate hospitality students (N = 205) concerning the measure of teaching feed from the college regarding interpersonal expertise with significant difference in mean of sampling data at 0.05 level.

Based on the above result, no "statistically-significant difference" in the scores of Table 9.10 (interpersonal expertise) about the perception of male students (M = 4.5, SD = 0.49) and female students (M = 4.4, SD = 0.52) was found. Null hypothesis is accepted.

TABLE 9.10   Importance of Result of Independent Sample t of Interpersonal Expertise

	Group	N	Mean	Standard Deviation	t	df	Sig. (2-Tailed)
Interpersonal	Male	127	4.5	0.49	0.66	203	0.51
Expertise	Female	78	4.4	0.52	0.67	169.17	0.50

Source: Primary data.

> ➢ **Hypothesis 10: H_0:** *There is no perceptual difference in male and female undergraduate hospitality students regarding the importance of Social Expertise.*

An independent–samples t-test was conducted to compare the perceptual difference of the male and female undergraduate hospitality students (N = 205) concerning the measure of teaching feed from the college regarding interpersonal expertise with significant difference in mean of sampling data at 0.05 level.

Based on the above result, no "statistically-significant difference" in the scores of Table 9.11 (social expertise) about the perception of male students (M = 4.5, SD = 0.48) and female students (M = 4.4, SD = 0.63) was found. Null hypothesis is accepted.

TABLE 9.11   Importance of Result of Independent Sample t of Social Expertise

	Group	N	Mean	Standard Deviation	t	df	Sig. (2-Tailed)
Social Expertise	Male	127	4.5	0.48	1.29	203	0.20
	Female	78	4.4	0.63	1.37	193.37	0.17

Source: Primary data.

9.5   CONCLUSION AND LIMITATIONS

The study of mean scores shows that there was a perceptual difference among male and female undergraduate hospitality students regarding the importance of employability expertise.

CHART 9.2   Importance of Employability Expertises.

Source: Primary data.

Similarly, the difference in perception among the male and female students has differed in perception on input level of employability expertise at the institutions.

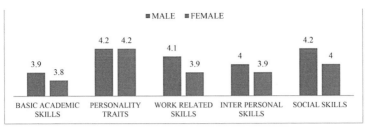

CHART 9.3   Level of Input at the Institutions.

Source: Primary data.

➢ Summary of Hypotheses:

Hypothesis No.	Hypothesis	Result
1	There is no perceptual difference between male and female undergraduate hospitality students regarding input by the institution on Basic Academic Expertise.	Accepted
2	There is no perceptual difference between male and female undergraduate hospitality students regarding input by the institution on Personality Traits.	Accepted
3	There is no perceptual difference between male and female undergraduate hospitality students regarding input by the institution on Work Related Expertise.	Accepted
4	There is no perceptual difference between male and female undergraduate hospitality students regarding input by the institution on Interpersonal Expertise.	Accepted
5	There is no perceptual difference between male and female undergraduate hospitality students regarding input by the institution on Social Expertise.	Accepted
6	There is no perceptual difference between male and female undergraduate hospitality students regarding the importance of Basic Academic Expertise.	Rejected
7	There is no perceptual difference between male and female undergraduate hospitality students regarding the importance of Personality Traits.	Rejected
8	There is no perceptual difference between male and female undergraduate hospitality students regarding of Work-Related Expertise.	Accepted
9	There is no perceptual difference between male and female undergraduate hospitality students regarding the importance of Interpersonal Expertise.	Accepted
10	There is no perceptual difference between male and female undergraduate hospitality students regarding the importance of Social Expertise.	Accepted

Source: Primary data.

As the sample size (N = 205) was not big enough, the above outputs could not be generalized for the undergraduate hospitality student association in the whole state. The other reason was that due to connectivity constraints in the pandemic situation, the researcher could reach only eight hospitality institutions from the state. The background reason for this perception among the students may be investigated in the future during further research. A comparative study may be conducted among different institutions in a state or city in India. A study can be conducted among the public and the private institutions. Similar research studies may be conducted in the future where the data from different Indian states can be collected.

KEYWORDS

- **data encryption standards**
- **employability expertise**
- **hospitality institutions**
- **optimal asymmetric encryption padding**
- **public-key cryptography**
- **secret-key cryptography**

REFERENCES

1. Christou, E. S., (1999). Hospitality management education in Greece an exploratory study. *Tourism Management, 20*(6), 683–691.
2. Millar, M., Mao, Z., & Moreo, P., (2010). Hospitality & tourism educators vs. the industry: A competency assessment. *Journal of Hospitality & Tourism Education, 22*(2), 38–50.
3. Lane, D., Puri, A., Cleverly, P., Wylie, R., & Rajan, A., (2000). *Employability: Bridging the Gap Between Rhetoric and Reality; Second Report: Employee's Perspective.* London: Create Consultancy/Professional Development Foundation.
4. Hillage, J., & Pollard, E., (1998). *Employability: Developing a Framework for Policy Analysis.* Research Brief, 85, ISBN 085522889 X, Nov. 1998.
5. Chapman, J. A., & Lovell, G., (2006). The competency model of hospitality service: Why it doesn't deliver. *International Journal of Contemporary Hospitality Management.*

6. Jauhari, V., (2006). Competencies for a career in the hospitality industry: An Indian perspective. *International Journal of Contemporary Hospitality Management.*

7. Raybould, M., & Wilkins, H., (2005). Over qualified and under experienced: Turning graduates into hospitality managers. *International Journal of Contemporary Hospitality Management.*

8. Munar, A. M., & Montaño, J. J., (2009). Generic competencies and tourism graduates. *Journal of Hospitality, Leisure, Sports and Tourism Education (Pre-2012), 8*(1), 70.

9. Agrawal, V., & Dasgupta, S., (2018). *Identifying The Key Employability Expertises: Evidence from Literature Review*, 85–90. Retrieved from www.iosrjournals.org (accessed on 11 October 2022).

10. Wickramasinghe, V., & Perera, L., (2010). Graduates,' university lecturers' and employers' perceptions towards employability expertise. *Education+ Training.*

11. Ju, S., Zhang, D., & Pacha, J., (2012). Employability expertise valued by employers as important for entry-level employees with and without disabilities. *Career Development and Transition for Exceptional Individuals, 35*(1), 29–38.

12. Aguila, G. M., De Castro, E. L., Dotong, C. I., & Laguador, J. M., (2016). Employability of computer engineering graduates from 2013 to 2015 in one private higher education institution in the Philippines. *Asia Pacific Journal of Education, Arts and Sciences, 3*(3), 48–54.

13. Oliveira, E. D. D., & Guimarães, I. D. C., (2009). *Employability Through Competencies and Curricular Innovation: A Portuguese Account.* Corpus ID: 41813441, October, 2009.

14. Dimalibot, G. A., Diokno, J. D., Icalla, M. F., Mangubat, M. R. C., & Villapando, L. C., (2014). Employment status of the tourism graduates of batch 2013 in lyceum of the Philippines University-Batangas. *Journal of Tourism and Hospitality Research, 11*(1), 46–56.

15. Boyadjieva, P., & Ilieva-Trichkova, P., (2015). Institutional diversity and graduate employability: The Bulgarian case. In: *Diversity and Excellence in Higher Education* (pp. 153–171). Brill Sense.

16. Dasgupta, & Agrawal, (2019). Hospitality stakeholders' perception on importance of employability expertise: Empirical evidence from West Bengal. *IJRAR, 6*(1), 161–166.

17. Li, F., Ding, X., & Morgan, W. J., (2009). Higher education and the starting wages of graduates in China. *International Journal of Educational Development, 29*, 374–381.

CHAPTER 10

Application of IoT and Big Data in Automated Healthcare

A. KRISHNARAJU,[1] K. SREENIVASAN,[2] P. MARISAMI,[3] B. PRAKASH,[4] and S. SRINIVASAN[5]

[1]*Professor, Department of Mechanical and Automation Engineering, PSN College of Engineering and Technology, Tirunelveli, Tamil Nadu, India*

[2]*Lecturer, Department of Production Engineering, Thiagarajar Polytechnic College, Salem, Tamil Nadu, India*

[3]*Lecturer, Department of Mechanical Engineering, Government Polytechnic College, Salem, Tamil Nadu, India*

[4]*Second Engineer, Synergy Marine Pt. Ltd., Singapore*

[5]*Deputy Manager, Product Engineering, Bharath Heavy Electrical Ltd., Trichy, Tamil Nadu, India*

ABSTRACT

Patient care excellence and safety, rapidly rising hardware costs and restrictions, healthcare costs, and increased computer speeds are some of these concerns. With improved patient information and real-time tracking provided by IoT, the chances of avoiding and curing diseases are greater than ever. A little but incredibly clever piece of digital technology is quickly challenging the conventional approaches to the use of pharmaceuticals in

System Design Using the Internet of Things with Deep Learning Applications.
Arpan Deyasi, Angsuman Sarkar, and Soumen Santra (Eds)
© 2024 Apple Academic Press, Inc. Co-published with CRC Press (Taylor & Francis)

medical treatments. The suitability of Bluetooth (BLE), Zigbee, and Wi-fi for use in advanced healthcare systems was then contrasted with short- and long-range communications. There are several benefits to assuming the e-health cloud will provide IoT solutions for the healthcare industry. To maintain scalable, individual, protected, and reliable operations for all manufacturing industries and medical in this chapter, there are many challenges that must be overcome. In this system with IoT, the problem is highlighted with emphasis on the significance concepts involved, implementations, applications structural, monitoring, and big data challenges.

10.1 INTRODUCTION

Internet of things (IoT) devices have in the development of opportunities and challenges and also additional important responsibilities, and on that condition those essential services such as venture companies, management departments, healthcare sectors and all public sectors [1–5]. Healthcare services have been aggressively using the IoT to betterment of the quality of patients' treatment, physicians, caretakers, and medical suppliers, all of which are dependent on application conditions. It is a very smart device that facilitates to the machine-machine and/or machine-human with an interfacing achievable organization system. An IoT system has been launched opportunities and challenges [6–10] such as gaming, finance, transportation, and healthcare [11–18]. IoT system has a global system that joins the device to the internet or Wi-Fi communications to each other using Wi-Fi sensors systems [19]. In this system, increasing technology has rapidly established that 60 billion gadgets will be ported to the globe by 2025. Several Smart applications will be developed in our systems, allowing us to be smarter, better, more effective, and more efficient while reducing unnecessary costs. In our innovation, the action in the health monitoring arrangement is one of the tough challenging due to the safety measures problems of data that know how to can be measured and also repeatedly improving and developing an innovative ideas or smart solutions to be find out to be updated the patients details and all the things. In this system, present a multiplicity of claim, and its benefits are observed to be extremely truthful and speedy recovery.

Here, data sharing has the following benefits: improving the efficiency, reliability, end to end description and usefulness of the proper service. In this system, many factors are to be measured, such as node, organization assessment, and response time range, high speed network coverage and also a model display. So, today, here an improved our system like in a technology growth such as interfacing communiqué linking the machines and mechanisms, humans, and data arrangement such as cloud, machine learning, and IoT. Work has previously has been in progress, and it is affecting the speed of thought in the various fields such as SMART City, Transportation, agriculture, healthcare, etc. Name of any fields and their applications under improvement of global innovations.

In our systems is a most vital part role of our life in all developed countries. Sadly, the steadily getting old population and the related rise in unending illnesses is placing significant nervous tension on modern healthcare systems, and the demand for resources from transports hospital beds to doctors and nurses are extremely high facilities without doctors' situations (like COVID-19). The IoT has been generally qualified as a possible solution to improve the demands on the healthcare systems in looks at monitoring patients with specific exclusively taking the all-diseases circumstances like patient's problems to be determined [35–37].

There are a demand and people are looking for young talent that knows this integration of sensors, embedded systems, and communication devices and their protocols. Computer science topics, especially data aggregation, data analysis, and the subsequent development of algorithms for decision making, are in high demand.

- The application layer is made up of wearable devices that can collect biomedical data such as moisture, temperature, pulse, brain, or heart electrical waves.
- The network layer stands on edge computing services such as data storage, signal processing, anonymization, and analysis in the cloud.
- The perception layer is dependent on the friendliness of the user. All communication is through Bluetooth, Zigbee or Wi-Fi and also sensors [19–26].

Non-invasive devices such as monitoring blood pressure, blood sugar and SpO2 healthcare devices at very low cost, wearable devices, real-time patient monitoring system within medical alert systems. Patients' body monitors the blood pressure, SpO2 and blood glucose level measured after feedback sends the doctor's phone.

In the global conditions systems are to be costlier depends upon the technology development, because of the increasing level of the population automatically at old aging people and (serpis) chronic disease also increasing in the developed countries. In all the country would go to supportive due to people would be more exaggerated in the chronic diseases [35–37].

10.2   INTERNET OF THINGS (IOT) APPLICATIONS

In recent years, extensive study has been undertaken on the advantages of cloud computing for healthcare applications [7, 18, 28, 31] as shown in Figure 10.1.

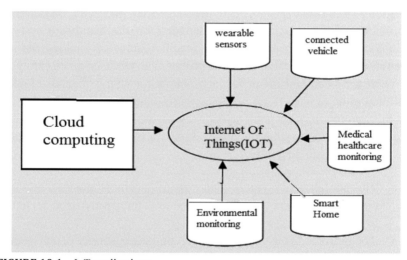

FIGURE 10.1   IoT applications.

10.2.1   ENVIRONMENTAL MONITORING

This system detects the prominence of our surroundings through an investigates the quality of utilization water such as intake liquids or bathing, all estimation of air moisture in the atmosphere conditions, earthquake finding, early forest fire identification, and other such things. It was connected remotely using IoT systems.

10.2.2   TRAFFIC MONITORING

In this system, the prominence of our global population surroundings quality of soil using road preparation and its constructions conditions, after GPS system using traffic jams on the road and its direction all things, and a that occur during functions, festivals, and other irregular problems are to be monitored.

10.2.3   AGRICULTURE

In this, monitor the prominence of our surroundings by analyzing plant sand and land procurement for harvesting or cropping to be maintained the raw seeds to final cropping monitored by the monsoon and atmosphere conditions and weather reports and all surroundings with the major help of sensors.

10.2.4   INSIDE HOUSES

This system can be observed and manage all electrical and mechanical and also communications devices like internet, television, smartphones and also save the water usage utilizing and electrical usage is possible control through by the systems.

10.2.5   HEALTHCARE SYSTEMS

In this system are suitable for popular rising innovations are fast developments for old aging persons, chronic illness patients not recovered, new medicines, and inventive wearable devices are continuously changing human service without disturbing the patient's feasibility. Technology will not be able to prevent the world's population from aging or developing chronic diseases. But some research developments also can't stop the area it is also going on progress improvements again and again.

10.2.6   PROBLEM IDENTIFICATION IN HEALTHCARE

In today's global humanity citizens are under a bundle of pressure to work nature and hard work pain in their jobs while also managing their families, the financial system, and the environment situations after end of his retired services are effected by chronic illness problems look like aged and or diseases along with the big hospitals is mainly due to shortage of beds or medicines or not adopted in the new technology to be occurred. Nowadays, the recent researchers are highly developed in day by day andand improve their healthcare technologies using smart techniques using in their system at high-speed internet communications in all essential fields. However, the success of IoT systems in our purpose for only remote area monitoring such as home, patients, physicians, hospitals, and wearable devices requiring short-range and long-range communications will be critical. In our systems, impartment building blocks are constructing an intellectual infrastructure to serve, manage, and support the devices that are increasing the world's population and save the time and save the life cycle.

The expanding performance of healthcare systems, as well as the appearance of big data in healthcare, are driving IoT support, advanced technology in an improved the wearing of the devices and Wi-Fi connectivity or Bluetooth, and growing infiltration of simultaneous devices in healthcare equipment. As a product, in the future scope of smart cities will be required smart hospitals in the future that can enable remote control of the flow of patients, provide critical situation, patient one-time access, actions, and analyze data to use the best approach for research or existing patents records are updated regularly. They will be improving their patients' outcomes and eminence or save the life for patients and also update the human data and medicines stored [41–45].

10.3   LITERATURE REVIEW

Silva et al. [6] have been described a nonstop growth in the IoT concept with consists of the world over computing, Wi-Fi, sensor networks, and machine-to-machine (M2M) interaction. It connects between mixed corporeal devices and communication among them via through the Internet. Moreover, highly widespread protocols and values are may obtain a comprehensive understanding of IoT summarized. However, IoT does not hold to a worldwide structural design.

Bhagya Nathali Silva [6] has described the architecture and related technologies forever rising idea with the new advancements of all places such as cloud computing, wearable sensor networks, and M2M interaction with the IoT. As a result, it describes widely recognized architectural concepts, which are further expanded with the associated communication protocols and standards.

Chauhan et al. [40] discussed an IoT, which a big revolutionary change, in which devices has a wide area in all fields are occupied such as the residence and civil structure of smart cities, computerization in industries, agriculture, individual's health for personalized health supervision and security threat in the healthcare industry.

Jagadeeswari et al. [32] have focused on recent emerging technologies in the medical internet of things and big data in personalized healthcare system significantly to fulfil the health check and assistive requirements of ageing people which all support various real-time appliances through improved services. The IoT in user data to discover the new message, predict before detection, build the decision over the dangerous circumstances for the progress of the quality of life. focused cloud, fog, big data computing and mobile-based applications.

Pradhan et al. [18] have studied the summary of the IoT-based technique great potential challenges in health care applications of interfacing various medical equipment such as wearable clothes, sensors, testing equipment virtual clinics all are to apply in remote places. That will improve patient safety, reduced cost, quality of services, increased or improve the patient health monitored. in the current aspects updates, the application of IoT based technology in future Healthcare Internet of Things (HioT) are solving numerous health care issues.

Riazul et al. [13] have reviewed the IoT in the global revolution of smart things in the critical structure blocks are an expansion of virtual invasive of all modern structure applications. The IoT has redesigning healthcare solutions with capable of global sustainable development in healthcare technologies perception so minimize security risk.

Keehyun et al. [35] have proposed the concept of an IoT systems are utilized in remote monitoring of the patients at home applications that are expanded to healthcare are applied a protocol system is a Multiclass Q-learning scheduling algorithm. It supports on the necessity of biomedical data delivery to medical staffs regularly.

Min-Woo et al. [42] have proposed a reliable M2M-based communication on the IoT system for personal healthcare device receiving the increasing awareness of the global population (patients) because they easily assist the remote monitoring of patients. The experiments show that the protocol conversion executes very efficiently.

Baker et al. [3] has challenged in a large amount concentration in IoT systems to modern living for it to be able to mitigate the harm inflicted by aging people on healthcare systems, as well as an increase in unrelieved disease. Aged peoples are healthier and Living longer life and also suggests us an incomparable Challenges in the condition of security, privacy, wearability, and low-power operation are all issues that the IoT must address. With the possibility of personal development, so far the current circumstances as a standardization live means preventive such as individual strength, weakness, and overall suitability for a wearable IoT healthcare systems.

Yin et al. [10] have been investigation of a multiplicity of in sequence of technologies (IoT) such as interacting and improve their strengthening in the existing healthcare services and old aged persons. In particularly applied to be connecting the medical resources and afford reliable, effective, and smart systems to old aged and disabled patients with a chronic illness.

Perrier [37] McKell institute has been observed two major work stoppage in the present of overcome the opportunities and challenges for the nation such as Automation and computerization in our global innovation likely replace entire industries surroundings and old aged person are affected by so many chronic diseases. Healthcare modernization can be improved productivity and fiscal sustainability. Living longer and healthier lives must be properly embattled to avoid vain care and waste.

Luigi et al. [12] have been addressed the main factors performed in the IoT through an integration of the significantly recent technologies and informatics telecommunications such as tracking, wired, Wi-Fi, sensors, and actuators.

Shancang et al. [16] has been explored in the global solutions of IoT in major parts are covered by heterogeneously connected all parts and virtual mechanism such as original structural design, expertise, and applications promising procedure for the execution, applications, and the major challenges.

Marko Kos et al. [31] have presented a new miniature wearable apparatus and an arrangement for video recording and detecting the group and biometric information such as worn on a wrist and can monitor covering temperature and pulse rate especially for swing-based sports applications like tennis or golf. Because of its device is powered by a light-weight Litho battery and miniature small size and minimum weight maximum performed by 6 hours of autonomy.

Sarkar and Misra [22] have developed new embedded systems as wireless body area network (WBAN) through microelectromechanical systems have concerned. In a global solutionm, such as enhancing the quality-of-life displacement form to an electrical signal.

Yin et al. [10] have been dedicated to global innovation applying different techniques such as information technologies (IT) in complement and growth in the existing healthcare systems. In the latest advanced technology, the IoT has been widely applied to all engineering, agriculture, and healthcare systems. Particularly in this healthcare systems interconnect the medical resources and reliable patients, effective, and smart healthcare service to the elderly and patients with a chronic illness.

Olaronke and Oluwaseun [8] has been examined the healthcare system consists of huge volumes of data which are frequently generated from various sources like physicians' daily case reports, hospital new case admission, patients discharge synopsis, medicals purchases, claim in insurance survey, medical imaging processing, lab reports, sensor-based devices, the study of whole genomes of organisms, social media articles in health check articles. Healthcare data, on the other hand, is extremely complicated and tough to maintain. The concept of the big data in benefits and attendant challenges such as ethical issues, usability issues as well as security and privacy issues are some of the factors considered the doing well achievement.

10.4 IOT-BASED HEALTHCARE SYSTEMS CONCEPT

IoT-enabled health-oriented systems involve various devices monitored by the basic concept, such as cloud computing, Wi-Fi, voice search and recognition, social media, 3D online and printing, smartphones, biometrics, magnetic resonance imaging (MRI) and electronic-based records or scans of health.

10.4.1 WIRELESS-BASED MOBILE TECHNOLOGY

This expertise would allow doctors to practice medicine from any point through any gadgets. It has an impact on almost every part of our lives. Tablets and smartphones are examples of mobile devices.

10.4.2 WEARABLE ORIENTED TECHNOLOGY

This understanding enables ordinary people to wear light-weight sensors dresses for patients and physicians' clothes, client protection and security protection for 24×7.

10.4.3   3-DIMENSIONAL PRINTING

This kind of print (3DP) is an advanced model process for generating 3-dimensional elements manufactured by all engineering assembling parts and gadgets, aerospace aviation, consumer items, medical healthcare items like wearable gadgets.

10.4.4   VIRTUAL REALITY (VR)

Virtual reality (VR) is an exceptionally intelligent, when they're away from their computers, they need to know about data protection and charging. It can help patients feel less anxious during medical-based procedures and safety situations.

10.4.5   ROBOTICS

Robots assume an important role in healthcare services in all places, such as surgery services without humans. They are filled with various medical needs to avoid human intervention.

10.4.6   CLOUD-BASED COMPUTING

Clouds computation creates the path manmade process suppliers convey administration for support less [34].

10.5   CLOUD-BASED IOT HEALTHCARE SYSTEMS

The survey on several kinds of literature, cloud technologies are mostly used Smart grid and mobile cloud computing for smartphones are examples of IoT applications. The potential of cloud technology for health record data is discussed, with a special emphasis on how a huge database may be utilized for data analysis and trend identification.

The following are benefits stem from the three primary services that can be provided by cloud technologies in healthcare environments:

1. **Software as a Service (SaaS):** Affords applications to healthcare that will allow them to interact with health data or complete other pertinent duties.

2. **Platform as a Service (PaaS):** It provides virtual tools for network, database management system, etc.

3. **Infrastructure as a Service (IaaS):** It supplies the physical infra-structure for storage, servers, and others.

10.6 LIST OF MANUFACTURERS IN THE IOT

The list of the IoT systems manufacturers and implementation benefits and also monitoring the systems from the key services that cloud technologies can deliver in healthcare facilities, such as given below:

- GE healthcare;
- Microsoft Corporation;
- SAP SE;
- Capsule Technologies, Inc.;
- Resideo Technologies, Inc.;
- Stanley Healthcare.

10.7 IMPLEMENTATION IN HEALTHCARE SYSTEMS

The following are implementation benefits from the major services that cloud technologies can provide in healthcare situations [34]:

- Low-cost methodology;
- Increased constant-monitoring and more caring for aging;
- The enhanced quality standardized of living-life;
- Increase in expectancy ratio of life;
- Reduce health-based material wastages;
- Improvement of better medicines and opportunities and research.

10.8 HEALTHCARE USED IN WEARABLE DEVICES

WBAN, which is known as wireless body area network, has been recognized as a solution component in the healthcare system based on IoT technology, such as the emergence of precise sensors with low performance structures is critical in this system. Mainly focused on the sensors are that non-obtrusive and non-invasive types of sensors such as implantable. Considered they are five fundamental sensors are used in the system. First three for monitoring the vital signs of beat, respiratory speed, and body heat, and a further two for monitoring blood pressure and blood oxygen, both commonly recorded in hospital environments [19–26].

10.8.1   PULSE SENSOR

It is an Arduino-based sensor to measure heart pulse rate.

10.8.2   RESPIRATORY RATE SENSORS

Respiratory rate sensors may be the most commonly used to read the pulse and is a vital indicator that can be used to detect a variety of emergency situations, including cardiac arrest, pulmonary embolism, and vasovagal syncope. Pulse sensors have been extensively studied for medical and fitness uses.

10.8.3   BODY TEMPERATURE SENSORS

Infra-Red (IR) temperature sensors facilitate an accurate non-contact temperature capacity in medical applications for this type measuring ear temperature, forehead temperature, or skin temperature.

10.8.4   BLOOD PRESSURE SENSOR

The non-invasive way of measuring blood pressure is done with the help of a blood pressure sensor. It's similar to a sphygmomanometer, but instead of a mercury column, it detects blood pressure using a pressure sensor.

10.8.5   PULSE OXIMETERS SENSORS

Pure Light sensor equipment uses high-quality LEDs and an adjusted receptor to abolish interference from minor frequencies.

10.8.6 OTHER WEARABLE SENSORS FOR HEALTHCARE

The sensors that developed several critical special health parameters are like heart health in wearable sensors that may be useful in the healthcare systems focused on monitoring a specific condition such as echo-cardio grams (ECG). To collect these signals, wearable sensors have been designed. Based on our research into the most cutting-edge technology in the field of wearable sensors.

- Focusing on healthcare monitoring system in the big data.
- Interactive health-based smart phone used to review apps.

10.9 HEALTHCARE IN COMMUNICATIONS

Communications are related to the IoT for healthcare can be classified into two main categories:

    i.   Short-range communications; and
  ii.   Long-range communications.

The previous communication used between devices within the WBAN, whilst the latter provides a connection between the central node of the WBAN and a base station. Both sorts of communications are regarded equally important in this chapter.

10.9.1 SHORT-RANGE COMMUNICATIONS

Short-range communications are used repeatedly between sensor nodes and the central node where data processing occurs in this wearable healthcare system, which can also be other goals, such as establishing mesh networks for smart lighting, could be pursued. Many short-range communications standards exist, but Bluetooth low energy (BLE) and Zigbee are likely to be the most widely utilized in IoT systems.

10.9.2 LONG-RANGE COMMUNICATIONS

In the urban context, long-range communications protocols with high applicability for IoT applications employed Low-Power Wide-Area Networks (LPWANs) with a range of several kilometers. This is far greater than the range of traditional IoT connectivity like Wi-Fi or Bluetooth. This device design permits for low-power, whose ranges are on the order of meters, making healthcare practicable would necessitate significant and pricey mesh networking or comparable. This decreases the chance of patients going offline and gives the wearer additional convenience. The most outstanding justification for broadcasting data from the central node to the Cloud storage is based on these advantages.

10.9.3   BLUETOOTH LOW ENERGY (BLE)

Low-energy Bluetooth (BLE) The special interest group (SIG) created Bluetooth to provide energy-efficient wearable gadgets which consist of a coin-cell battery-operated devices such as smartphones. It also connects with small peripheral and processing devices that enable IoT which is suitable for healthcare applications.

10.9.4   ZIGBEE

It was created by the Zigbee Alliance Company with the sole purpose of creating low-cost, low-power M2M networks based interconnections. Dissimilar types of modules supply and different characteristics based on the physical standard IEEE 802.15.4.

10.9.5   WI-FI

The Wi-Fi signals are taken out from advantageous off-the-shelf smartphones and laptops. Initial achievement of Wi-Fi-based human being activity sensing but to utilize the signals to automatically recognize the user's diet moments, some challenges require to be determined such as data-rate, band, specification, coverage, nodes, mechanism, protection, battery, and also cost of the systems. Comparison of short-range communication standards Bluetooth, Zigbee, and Wi-Fi and their comparison is already reported [26].

Overall, healthcare systems in Zigbee are almost compatible with numerous security mechanisms range, data rate, and power consumption for certain applications, and discovered that low-power, high-data-rate modules with an acceptable range were also available. The key disadvantages of Zigbee communications are that explanation might substitute cooperation unless the manufacturer implements it exceptionally well such as Medtronic PLC, Koninklijke Philips.

10.10   BIG DATA IN HEALTHCARE, CHALLENGES, AND RESOLUTIONS

10.10.1   DATA COLLECTED

Multiple data types can be collected and stored in IoT Healthcare systems, mainly focusing and monitoring on the dynamics of your health systems and the wearable devices also environments to improve the data quality [30–33]. The devices are stored in the assure servers or cloud database are major classifications in two types:

   i. SQL; and
  ii. NOSQL.

10.10.2   PROTECTION OF IOT-BASED SYSTEMS

- Cyber security is the significant challenge in a right measure in security concerns, and issues of the systems.
- Ensure end-to-end security system.
- The dealer must explanation of the device and reflect on the measurements as required for both sensors and their applications.

10.10.3   REAL-TIME REPORTING AND MONITORING

- That can save human lives in the time of medical urgent like heart failure, diabetes, asthma, and bone fracture, etc. [35–37].
- IoT in real monitoring in the situation in place, a smart medical installment connected to a Smartphone app.
- Healthcare application in IoT to maintaining or monitoring old aging people required data collection and connection of the smartphone to a physician directly solve the problem.

- Here patient details connected with healthcare 50% reduction in 30 days.
- In our device, collects and transfer the health data such as oxygen and blood sugar levels, blood pressure, and ECGs from all human bodies testing exclusively old aging and also unconscious patients.
- All the data are stored in the cloud and can be shared with an authorized physician, or family doctor or your insurance company, or an external specialist to allow them at the collected data regardless of their place, time, or opaque.

10.10.4   END-TO-END CONNECTIVITY

It enables Machine-Machine (M2M) communication information exchange-data management makes service effective by using protocols such as Wi-Fi, Bluetooth, z-wave, zig bee and other protocols, etc. [19].

Mainly technology development in the IoT service setup brings down the medical cost and immensely visits physician cutting downtime waste and also utilize better quality resources, improve the allocation planning.

10.10.5   PREDICTING THE NUMBER OF PATIENTS IN HOSPITALS USING DATA ANALYTICS

One of the biggest issues is deploying the appropriate quantity of people at the correct time explicit times but admissions patients to the hospital do not have a similar number of few hospitals to a review, BDA to handle the problem of staffing, both in terms of cost optimization and quality improvement patients reduce the waiting times their customer service.

10.10.6   PREDICTING SEPSIS RISK AND DEATH USING DATA ANALYTICS RESEARCHERS

Researchers are continuous improvements for admitted patients have routine health information such as various testing reports such as heart pulse rate, blood pressure, temperature, sugar, and the number of white blood cells discovered in electronic-health records (EHRs).

10.10.7 PREDICTING READMISSIONS IN HEART FAILURE PATIENTS USING DATA ANALYTICS

Researchers are continuous improvements for admitted within 30 days of discharge, the patient is at risk of being readmitted to the hospital. All patient data from a hospital's electronic health record (EHR) was de-identified report.

10.10.8 USING DATA ANALYTICS TO PREDICT THE DECAY

Researchers are continuous improvements for admitted patients affected by decay to using end-tidal monitoring from the home itself. Some of these existing monitoring systems such as electrocardiographic pulse monitoring and oxygen observed the patients for many years.

10.11 CHALLENGES IN HEALTH DATA ANALYTICS

- The vast amount of data is sent in a very short time application;
- Collect, report, and analyzes the data in real-time application;
- Tracking and alerts;
- One time alert is critical in life-threatening circumstances, medical IoT;
- IoT device gather vital data and transfer the data of doctors for real-time tracking.

10.11.1 INFRASTRUCTURAL CHALLENGES

- High-speed internet and latest equipment installed;
- High confidentiality, honesty, and accountability of data;
- On-demand communications provisioning;
- Expectation environment for data storage and data processing confidentiality and auditing capabilities;
- Provide an end-to-end source verification.

10.11.2   SECURITY AND PRIVACY CHALLENGES IN E-HEALTH CLOUD

Normally, the e-healthcare data should be in guarded security [39, 40] and privacy protection of patients' records domains:

i.   **Privacy:** Make sure that data is unauthorized access.
ii. **Honesty:** Ensuring the accuracy and reliability of data.
iii. **Endorsement:** Guarantee the nominee.
iv. **Isolation:** Guarantee collection about data.
v.   **Access Control:** Guarantee access levels.
vi. **Audit:** Ensuring the safety of data and the e-Health purpose.

10.11.3   BENEFITS OF IOT FOR HEALTHCARE

These are following are implementation from the major services that cloud technologies can provide in healthcare situations collecting patients:

- High quality of devices for improved the patient care updated;
- Reduced cost for re-admission and also easy paperless work;
- Explain the issue of resources scarcity;
- Maintain the patients all report and research encouragement;
- Support to national security levels, and new strategic planning;
- Support clinical test and financial operations;
- Facilitate form registers;
- High data security risks, loss of data and systems unavailability.

10.11.4   EMERGING TECHNOLOGY IN HEALTHCARE

In our emerging technological healthcare systems [28, 29, 38] there is a vital role in the IoT. Every process is kept systematically to the approach in the prevention of major factors to be measured [46]:

1. **Virtual Clinics:** mean video conferencing supplemented by biometrics and remote monitoring allows the participants to feel the feedback if any problem diagnostics or room facilities or improve the medical supply such as patients or physician or

caretaker or medicine supplier or insurance surveyor or hospital management, etc.

2. **Wearable:** integration interacts with using customer or consumer (such as patients) wearable technology to supplement healthcare, such as sensors to facilitate observing daily health patients' routine life; additionally, his report sent feedback to the hospital.

3. **Discharge:** summary report and also improvements in hospital at home, such as 24×7 access to watches remotely.

4. **Mobility:** supporting is advance in healthcare systems and solutions such as a patient bed that converts into a wheelchair or depends upon the needs used in the reconfigurable mechanism or robot.

5. **Informatics:** It is processing and planning the schedule for professionals like physician, such as a smart mobile phone that always keeps updating the health information from the patient to hospital or physician or caretaker.

10.12   RESULTS AND DISCUSSION

In this system is a unique and innovative characteristic of the IoT, a lot of challenges need to be overcome to preserve scalable, individual, protected, and reliable operations for all manufacturing industries and medical in this chapter. In this system with IoT identified the problem with importance on the significance concepts involved, implementations applications structural, monitoring, and big data challenges.

Mostly all things are equipped with sensors with actuators entering the global market population using a huge collection of technologies incorporated into the devices through management of things become extremely challenging operations. We emphasize the different facets of environments that contribute to aged patients and chronic illness patients and also diabetes report in monitoring, automatically send the feedback to hospital and all. The healthcare system is a huge effort and offers, secure, and low-cost IT solutions. Further continuous researchers [41–46] put efforts should be made to improve the sensor's degree of excellence until they are highly reliable, accurate, and comfortable to wear devices.

10.13   CONCLUSION

Modern research utilizes new cloud technologies applied for data storage and the most excellent performance for collecting and organizing big data in healthcare systems. In particular, we reviewed the investigations into the most advanced model of IoT-based healthcare systems in fields such as wearable sensors, interactions data standards, and cloud data storage methods, which can be applied to challenging tasks and monitoring the patients and physician-specific conditions and hospital observing, as well as all feedback.

These challenges consist of patient care superiority and safety, dramatically growing hardware costs and limitations, healthcare expenses, advanced computing speeds. Short and the suitability of long-range communications for advanced healthcare was then compared with systems are observed through Bluetooth (BLE), Zigbee, and Wi-Fi. Assumes the e-Health Cloud to supply IoT solutions for healthcare comes with many advantages:

- Growing the supply, scalability, and adaptability of the health information systems;
- Better patient care monitoring and also feedback from all environments;
- Data are confidentiality, integrity, and accountability;
- The investigation goes into the technologies within the fields of wearable sensors, interactions standards, and cloud data storage methods.

10.14   RECOMMENDATIONS FOR FUTURE WORKS

The survey validates that the IoT systems are implemented in various areas of healthcare and also further research is going on. In terms of sensors, significant progress has been made, but more progress is essential, particularly in the accuracy of complex devices such as blood pressure and respiratory rate sensors, both of which would be very useful to the field for easy diagnosis and provide treatment sooner. As such, further research efforts should be made towards improving the quality of excellence in these sensors until they are greatly comfortably wearable devices. This highlights the need for end-to-end security and privacy protection solutions, such as a guarantee that things are accessed only by authorized entities securely.

KEYWORDS

- **cloud computing**
- **healthcare**
- **internet of things**
- **oral communication**
- **security**
- **smartphones**
- **wearable devices**
- **Wi-Fi**

REFERENCES

1. Abu, K. E., Mohamed, N., & Al-Jaroodi, J., (2012). e-health cloud: Opportunities and challenges. *Future Internet, 4*, 621–645. https://doi.org/10.3390/fi4030621
2. Nižeti´c, S., Šoli´c, P., López-De-Ipiña González-De-Artaza, D., & Patrono, L., (2020). Internet of things (IoT): Opportunities, issues and challenges towards a smart and sustainable future. *J. Clean. Prod., 274*, 122877.
3. Baker, S. B., Xiang, W., & Atkinson, I., (2017). Internet of things for smart healthcare: Technologies, challenges, and opportunities. *IEEE Access, 5*, 26521–26544. doi: 10.1109/access.2017.2775180.
4. Pawan, S., (2018). *Internet of Things Based Health Monitoring System: Opportunities and Challenges 9*(1). doi: http://dx.doi.org/10.26483/ijarcs.v9i1.5308.
5. Eman, A. K., Nader, M., & Al-Jaroodi, J., (2012). e-health cloud: Opportunities and challenges. *Future Internet, 4*, 621–645. doi: 10.3390/fi4030621.
6. Bhagya, N. S., Murad, K., & Kijun, H., (2017). Internet of things: A comprehensive review of enabling technologies, architecture, and challenges. *IETE Technical Review*. doi: 10.1080/02564602.2016.1276416.

7. Sweta, A., Anil, K. G., & Mohammad, D., (2020). Challenges of IoT in healthcare. *IoT and ICT for Healthcare Applications, Book Series: Springer Innovations in Communication and Computing,* ISBN: 978-3-030-42934-8.

8. Olaronke, & Oluwaseun, O., (2016). Big data in healthcare: Prospects challenges and resolutions. *Proc. Future Technol. Conference. (FTC)*, 1152–1157 doi: 10.1109/ftc.2016.7821747.

9. Ahad, M. T., & Yau, K. L. A., (2019). 5G-based smart healthcare network: Architecture, taxonomy, challenges and future research directions. *IEEE Access, 7*, 100747–100762. doi: 10.1109/ACCESS.2019.2930628.

10. Yin, Y., Zeng, Y., Chen, X., & Fan, Y., (2016). The internet of things in healthcare: An overview. *Journal of Industrial Information Integration, 1*, 3–13. doi: 10.1016/j.jii.2016.03.004.

11. Shancang, L., Li Da, X., & Shanshan, Z., (2015). The internet of things: A survey. *Information Systems Frontiers, 17,* 243–259.

12. Luigi, A., Antonio, I., & Giacomo, M., (2010). The internet of things: A survey. *Computer Networks, 54,* 2787–2805. doi: 10.1016/J.Comment.2010.05.010.

13. Riazul, I. S. M., Daehan, K., Md Humaun, K., Mahmud, H., & Kyung-Sup, K., (2015). *The Internet of Things for Health Care: A Comprehensive Survey, 4.*

14. Yin, Y., Zeng, Y., Chen, X., & Fan, Y., (2016). The internet of things in healthcare: An overview. *J. Ind. Inf. Integr., 1,* 3–13 https://doi.org/10.1016/j.jii.2016.03.004.

15. Birje, N., & Hanji, S. S., (2020). Internet of things based distributed healthcare systems: A review. *Journal of Data, Information and Management, 2,* https://doi.org/10.1007/s42488-020-00027-x.

16. Shancang, L., Li Da, X., & Shanshan, Z., (2014). *The Internet of Things: A Survey, Volume 26.* doi: 10.1007/s10796-014-9492-7.

17. Verdejo, E. Á., López, J. L., Mata, M. F., & Estevez, M. E., (2021). Application of IoT in healthcare: Keys to implementation of the sustainable development goals. *Sensors, 21,* 2330. https://doi.org/10.3390/s21072330.

18. Pradhan, B., Bhattacharyya, S., & Pal, K., (2021). IoT-based applications in healthcare devices. *Journal of Healthcare Engineering,* 6632599. https://doi.org/10.1155/2021/6632599.

19. Chu, M., Nguyen, T., Pandey, V., et al., (2019). Respiration rate and volume measurements using wearable strain sensors. *NPJ Digital Med., 2,* 8. https://doi.org/10.1038/s41746-019-0083-3.

20. Gope, P., & Hwang, T., (2016). BSN-care: A secure IoT-based modern healthcare system using a body sensor network. *IEEE Sensors J., 16,* 1368–1376 doi: 10.1109/JSEN.2015.2502401.

21. Yang, X., (2015). Textile fiber optic microbend sensor used for heartbeat and respiration monitoring. *IEEE Sensors J., 15*(2), 757–761 doi: 10.1109/JSEN.2014.2353640.

22. Sarkar, S., & Misra, S., (2016). From micro to nano: The evolution of wireless sensor-based health care. *IEEE Pulse, 7*(1), 21–25. doi:10.1109/MPUL.2015.2498498

23. Matin, M. A., & Islam, M. M., (2012). Overview of wireless sensor network. *Wireless Sensor Networks - Technology and Protocols.* doi: 10.5772/49376.

24. Von, R. W., Chanwimalueang, T., Goverdovsky, V., Looney, D., Sharp, D., & Mandic, D. P., (2016). Smart helmet: Wearable multichannel ECG and EEG. *IEEE J. Transl. Eng. Health Med., 4,* doi: 10.1109/JTEHM.2016.2609927.

25. Michael, C., Thao, N., Vaibhav, P., Yongxiao, Z., Hoang, N. P., Bar-Yoseph, R., Radom-Aizik, S., et al., (2019). Respiration rate and volume measurements using wearable strain sensors. *NPJ Digital Medicine, 2*(8), 9. https://doi.org/10.1038/s41746-019-0083-3.

26. Md Abbas, A. K., & Md Alamgir, K., (2016). *Comparison Among Short-Range Wireless Networks: Bluetooth, Zigbee, & Wi-Fi* (Vol. 11, No. 1). Daffodil International University Library. http://hdl.handle.net/20.500.11948/1466 (accessed on 11 October 2022).

27. Fan, Y. J., Yin, Y. H., Xu, L. D., Zeng, Y., & Wu, F., (2014). IoT-based smart rehabilitation system. *IEEE Trans. Ind. Information, 10*(2), 1568–1577. doi 10.1109/ACCESS.2015.2437951.

28. Matthew, N. O. S., Nishu, G., Yogita, P. A., & Sarhan, M. M., (2020). Emerging IoT technologies in smart healthcare. *IoT and ICT for Healthcare Applications*, 3–10.

29. Pasluosta, C. F., Gassner, H., Winkler, J., Klucken, J., & Eskofier, B. M., (2015). An emerging era in the management of Parkinson's disease: Wearable technologies and the internet of things. *IEEE J.*

30. Sabyasachi, D., Sushil, K. S., Mohit, S., & Sandeep, K., (2019). *Big Data in Healthcare: Management, Analysis and Prospects, Future.* https://doi.org/10.1186/s40537-019-0217-0.

31. Marko, K., & Iztok, K., (2017). A wearable device and system for movement and biometric data acquisition for sports applications. *IEEE Access, 1*, 99. doi: 10.1109/access.2017.2675538.

32. Jagadeeswari, V., (2018). A study on medical internet of things and big data in the personalized healthcare system. *Health Information Science and Systems, 6*, 14, https://doi.org/10.1007/s13755-018-0049-x.

33. Shanmugasundaram, G., & Sankarikaarguzhali, G., (2017). An investigation on IoT healthcare analytics. *International Journal of Information Engineering and Electronic Business, 9*(2), 11–19. doi: 10.5815/ijieeb.2017.02.02.

34. Khan, A. R., Othman, M., Madani, S. A., & Khan, S. U., (2014). A survey of mobile cloud computing application models. *IEEE Commun Surveys Tuts., 16*(1), 393–413. doi: 10.1109 /SURV.2013.062613.00160.

35. KeeHyun, Park, J. P., & JongWhi, L., (2017). An IoT system for remote monitoring of patients at home engineering. *Applied Sciences.* https://doi.org/10.3390/app7030260.

36. Kadhim, T. K., Ali, M. A., Salim, M. W., & Hussein, T. K., (2020). An overview of patient's health status monitoring system based on internet of things (IoT). *Wireless Personal Communications, 114,* 2235–2262.

37. Perrier, E., (2015). *Positive Disruption: Healthcare Ageing and Participation in the Age of Technology.* Sydney, NSW, Australia: The McKell Institute. https://apo.org.au/node/57042 (accessed on 11 October 2022).

38. Bajwa, M., (2014). Emerging 21st-century medical technologies. *Pakistan J. Med. Sci., 30*(3), 649–655.

39. Yuchen, Y., Longfei, W., Guisheng, Y., & Lijie, L., (2017). A survey on security and privacy issues in internet-of-things. *IEEE Internet Of Things Journal, 99,* 1250–1258. doi: 10.1109/JIOT.2017.2694844

40. Chauhan, Y. H., (2020). *Personalized Health Monitoring Using Machine Learning and Blockchain for Security of Data, Security and Trust Issues in the Internet of Things*. IoT-based healthcare. ISBN: 9781003121664.

41. Gubbi, J., Buyya, R., Marusic, S., & Palaniswami, M., (2013). Internet of things (IoT): A vision, architectural elements, and future directions. *Future Generation Computer Systems, 29*(7), 1645–1660.

42. Min-Woo, W., JongWhi, L., & KeeHyun, P., (2018). A reliable IoT system for personal healthcare device. *Computer Science Future Gener. Comput. Syst.* doi: 10.1016/j.future.2017.04.004.

43. Majumder, S., Rahman, M. A., Islam, M. S., & Ghosh, D., (2018). Design and implementation of a wireless health monitoring system for remotely located patients. In: *2018 4th International Conference on Electrical Engineering and Information & Communication Technology*. doi: 10.1109/ceeict.2018.8628077.

44. Mahmoud, E., & Hon, S. C., (2016). Internet of things applications: Current and future development. *Innovative Research and Applications in Next-Generation High-Performance Computing,* 397–427. doi: 10.4018/978-1-5225-0287-6.ch016

45. Bibri, S. E., & Krogstie, J., (2017). Smart sustainable cities of the future: An extensive interdisciplinary literature review. *Sustain. Cities Soc., 31,* 183–212. doi: 10.1016/j.scs.2017.12.034.

46. Darwish, A., Hassanien, A. E., Elhoseny, M., Sangaiah, A. K., & Muhammad, K., (2017). The impact of the hybrid platform of internet of things and cloud computing on healthcare systems: Opportunities, challenges, and open problems. *Journal of Ambient Intelligence and Humanized Computing.* doi: 10.1007/s12652-017-0659-1.

47. https://www.finoit.com/iot-application-development-company/ (accessed on 11 October 2022).

CHAPTER 11

Forecasting the Impact of COVID-19: Data Analysis

SHALINI ADHYA,[1] SANGHITA KUNDU,[2] and MAINAK BISWAS[3]

[1]*Arya Parishad Vidyalaya for Girls, Kolkata, West Bengal, India*

[2]*Department of Education, Institute of Education for Women, Hastings House, Kolkata, West Bengal, India*

[3]*Department of Electrical Engineering, Techno International New Town, Kolkata, West Bengal, India*

ABSTRACT

The manuscript deals with forecasting COVID-19 analysis based on collected data and corresponding analysis. The heterogeneity of data is maintained by considering several classes for input parameters, along with consideration of the post-lockdown period. Through case studies, the impact of Covid on other diseases is investigated, along with job status, online teaching-learning, etc. The effect of 2nd wave is elaborately discussed in a separate section, which helps to investigate the impact of Covid 19 in detail. The data analysis and data forecasting make it abundantly evident that COVID-19 is becoming less widespread every day while the likelihood of obtaining the vaccination is gradually rising. One of the biggest issues is the population. In addition, the use of technology, IT, and online teaching and learning is quickly expanding in the modern world.

System Design Using the Internet of Things with Deep Learning Applications.
Arpan Deyasi, Angsuman Sarkar, and Soumen Santra (Eds)
© 2024 Apple Academic Press, Inc. Co-published with CRC Press (Taylor & Francis)

11.1 INTRODUCTION

COVID-19 is a coronavirus infection produced by a family of corona-viruses that causes illnesses such as the common cold. Beginning in the year 2019, it is producing Severe Acute Respiratory Disease (SARS) and Middle East respiratory syndrome (MERS) [1] by altering its variations and showing a wide range of symptoms. It has said to be initially originated in China in and around November 2019 with the onset of the first case in the market of Wuhan. The World Health Organization declared it a Public Health Emergency of International Concern on January 30, 2020, and subsequently declared it a pandemic on March 11, 2020 [2], as the disease's emission was accumulating at an unfathomable rate and causing millions of deaths throughout the world. On May 22, 2020, 166 million cases had been verified over the world, with 3.4 million fatalities. The virus that causes severe acute respiratory syndrome (SARS-CoV-2) is a member of the coronavirus family, which includes viruses found in bats and pangolins [2]. The virus can be detected by taking a swab from the nose and throat under medical circumstances and detecting whether or not the virus is present. Coming in contact with an infected person can cause the other person to contaminate the disease. When a COVID infected person sneezes, coughs, talks, or shares food, the worst virus may be spread fast and easily from one person to another. The infected person's droplets do not hover in the air but rather recline on the ground since they are too heavy to float in the air. These droplets can enter the body when a person inhales, touches any surface which consists of the virus, and touch the face or eyes. Here, two works are done simultaneously, one is the data analysis part of the dangerous biological weapon Coronavirus and on the other hand, the next part is the impact of COVID-19 are discussed and the scenario is forecasted using the exponential algorithm [14], where it is very clear that the condition will be changed in no time. But for that the need for vaccination is very important, this is the challenge nowadays.

11.2 STATUS OF COVID-19: DATA ANALYSIS

According to recent findings, it primarily affects people within a 6-foot radius of each other [2]. During the early days of the disease, it was unknown to the doctors and scientists how it was spreading. Figure 11.1

shows the data for COVID-19 spread from February to May 2020. Figures 11.2 (according to the research of 2020) and 11.3 (for children) show the symptoms at a glance [11].

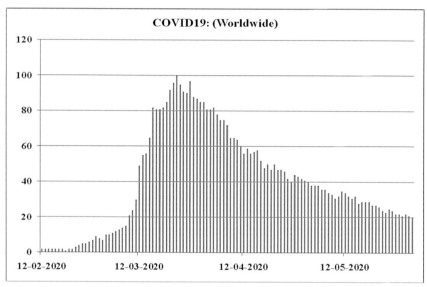

FIGURE 11.1 COVID-19 spread from February to May 2020.

Fever, cough, loss of taste and smell, tiredness, shortness of breath or difficulty breathing, muscle pains throughout the body, chills, sore throat and runny nose, headaches, chest pain (mild-severe), conjunctivitis (rare), nausea, vomiting, diarrhea, rashes are the symptoms of the virus that individuals should.

The range of its contamination can be calculated from mild to severe, showing different symptoms as per individuals. Some might show all the above-mentioned symptoms whereas others might show few of them. Some people were asymptomatic but were able to contaminate the other person if the other person touches the contaminated surfaces and touched his eyes, nose, or face making him COVID positive. The asymptomatic person may not be at higher risk but by transmitting the disease to others who might be suffering from an autoimmune disease or might be a patient of cancer, diabetes, hypertension, asthma, pregnancy, etc. These patients are always at high risk than the others. Complications among such patients may lead to clotting of blood, pneumonia, organ failure, dysfunctional kidneys, etc.

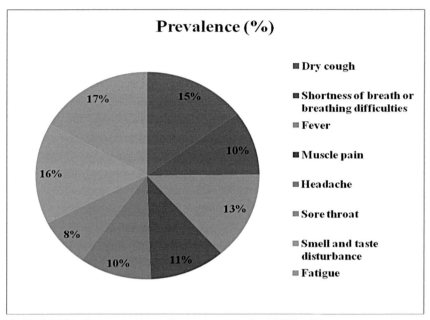

FIGURE 11.2   Symptoms (according to the research of 2020) with percentage.

The only way to stop the spread is to take preventative steps like sanitization and wearing masks. People, institutions, society, and local and national governments have documented measures such as separating from the public at home and outdoors, sanitizing, and wearing a mask in public. To track the spread of the disease and take appropriate precautions, measures such as detecting symptoms and isolating individuals, tracing infected people and quarantining them, social and physical distancing measures such as mass gatherings, international and national travel measures, and vaccines and treatments [3] were calculated.

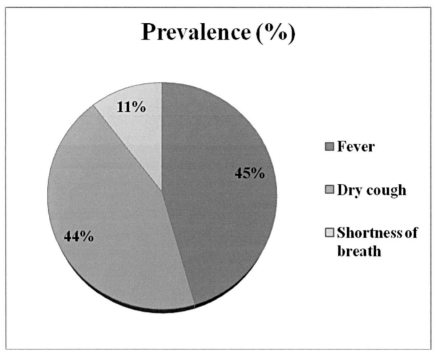

FIGURE 11.3   Symptoms (children) with percentage.

11.3   LOCKDOWN AND UNLOCK

The measure of Social and physical distancing helped to slow the spread of disease by stopping chains of transmission of COVID-19 and preventing new ones from transmitting. Lockdown was announced in some parts of the world to slow the transmission level to the greatest extend. Sanitizing roads and restaurants so that the contaminated areas can be cleaned, families were asked to isolate themselves from each other to stop the chain from increasing. Avoiding over-crowding, visits to malls, and shopping were among some of the measures taken. Work from home and teleconferencing, e-commerce, and e-learning were made compulsory as all the offices, schools, and public places were shut for a few months. The rise in cases all over the world forced people to follow the rules and regulations. Awareness regarding the spread of COVID-19 spread throughout the world spreading words of masking themselves while in public, cough etiquette and social distancing became a norm in normal.

During the outbreak of the disease, the ultimate solution, according to the Government was an announced lockdown for 21 days. So, the Central Government announced a lockdown for 21 days on 22nd March 2020 to check the transmission and further steps would be taken by the situation. Figure 11.4 shows the lockdown dates with days.

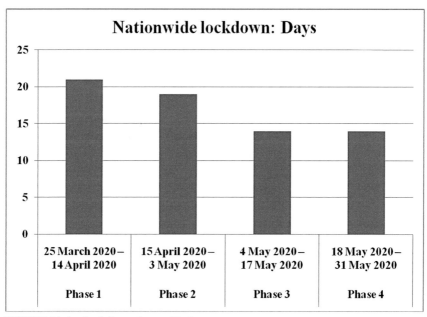

FIGURE 11.4   Lockdown dates with days.

It would have been easier to contain the pandemic if India had not been the world's second most populous country. Individual mobility restrictions were introduced as soon as people began to be more optimistic and social separation became the norm. As of April 7, 2020, around 5,000 confirmed cases had been recorded in India, with 90% of patients being healed. Before the epidemic engulfed the globe, the rate of mortality is 03% [8]. The Government had come up with different strategies for helping the poor and needy in this situation. Due to the lockdown and upturned situation of the economy, the struggle of the majority of the citizens was in a thrash which was somehow taken care of by the Government and also a lot of NGOs which took part in controlling the situation. After that few restrictions were waved, and these phases are called unlock, Figure 11.5 shows the dates along with the days of the phases.

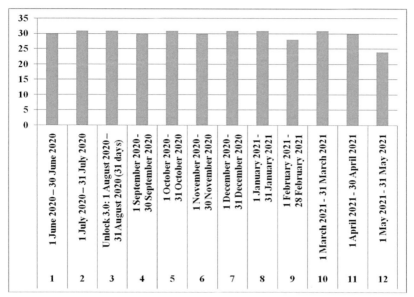

FIGURE 11.5 Unlock phase.

Around October, there was a sharp drop in the COVID cases and it was a sigh of relief for the Indians and brought a new ray of hope among Indians. There were trials of vaccine tests but was no confirmation of vaccines were to be effective.

11.4 CASE STUDY

According to the Clinical Management Protocol for the New Coronavirus, which was issued by the Brazilian National Ministry of Health in February this year, males and people over the age of 50 predominated among the first 99 patients brought to a hospital in Wuhan with pneumonia and a confirmed diagnosis of COVID-19. The analysis is depicted in Figure 11.6. Lymphopenia was detected in a separate study including 41 people who had been diagnosed with COVID-19. Guan et al. [4] looked at studied data from 1,099 patients in China who had confirmed COVID-19 and found Guan et al. [4] analyzed data from 1,099 patients in China who had confirmed COVID-19 and discovered that the average age of the patients was 47 years old, with 41.9% of the patients being female. In 67 individuals, the predominant combined consequence was admission to the intensive care unit (ICU), the need for mechanical breathing, or death (6.1%).

The average incubation time was four days. The majority of patients did not have a temperature and had normal X-ray results. Based on a survey of 55,924 verified cases, the World Health Organization-China Joint Mission on Coronavirus Disease 2019 [3] reported the most common signs and symptoms: fever, fatigue, and vomiting [4]. About 80% of individuals with COVID-19 who had it verified in the lab had mild to severe symptoms, which comprised both pneumonia and non-pneumonia instances. Dyspnea, a respiratory rate of less than 30 breaths per minute, a peripheral oxygen saturation level of less than 93%, and an arterial oxygen tension/fraction of oxygen ratio of less than 300 were all symptoms indicating severe sickness in 13.8% of the participants. In the first 24 to 48 hours, 6.1% experienced a severe disease, which included respiratory catastrophe and septic jolt, with or without multiple organ dysfunction/failure; Although asymptomatic infection has been documented, the percentage of people who are asymptomatic is unclear.

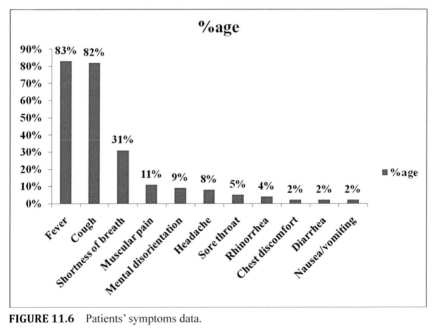

FIGURE 11.6   Patients' symptoms data.

Ones over 60 years old, especially those with underlying conditions such as hypertension, diabetes, cardiovascular disease, chronic respiratory disease, and cancer, are stated to be the people most at risk of serious illness and death [4].

After becoming infected with the virus, the incubation period ranges from 2 to 14 days. Symptoms of the virus included a dry cough, fever, and/or loss of smell and taste, as well as a painful throat that ranged from mild to severe. It was suggested that anyone who had direct touch with someone who tested positive with the virus separate themselves from others, particularly the elderly. Even if a patient's test results are positive, he or she does not need to be admitted to the hospital; instead, patients can recuperate at home with the help of recommended medication and a well-balanced diet.

11.5   COVID-19 IN INDIA

While the world has counted 16.6 crores [5] (only medically tested till 23.05.2021) of tested coronavirus patients, India has tested 2.3 Crores (as tested as and on 23/04/2021) positive cases of coronavirus. On January 27, 2020, the first case of COVID-19 infection was reported in Kerala. A 20-year-old returnee from Wuhan, China, and a student at Wuhan University went to the hospital on January 23rd for a test and remained asymptomatic until January 26th, when she complained of a dry cough and sore throat the next morning. On the 27th of January, she awakened with a little sore throat and a dry, persistent cough. According to the girl, there was no history of interaction with a person who was suspected or confirmed to have a COVID-19 illness. She went on to say that though she had not visited the Wuhan wholesale market, she had come into contact with persons who were exhibiting symptoms. As soon as she returned from Wuhan, China, she received instructions from the Kerala State Authorities and Government to quarantine herself and monitor her symptoms [6]. She was kept under continual surveillance for days, including swab testing, oxygen level measurements, and body temperature checks. Her sample was collected for 14 days in a row, and the results were positive; On the 17th day, the negative result was confirmed. It was unclear how long it took the body to produce anti-body, and positive instances of the virus began spreading more quickly in India.

The Indian government decided to vacate over 250 Indians from Wuhan, China where the pandemic has killed many lives. The Union health ministry had taken steps to request the returnees of China to isolate themselves and self-report themselves in case of any COVID-19 symptoms so that immediate action can be taken for them and the chain of spread can be stopped [7].

On 22nd February the Government imposed restrictions on movements and the region was divided into two – red zones and yellow zones. Red zones included areas like Lombardy and Veneto, and outside of these regions were marked as yellow zones. Schools and municipalities in and around this area were closed. All the educational trips were canceled and people were asked to buy supplies for at least 14 days [10].

All the community services, matches as well as pre-dated Carnivals were canceled or postponed for few days but later canceled by Prime Minister Giuseppe Conte. To halt the virus's spread, health officials established emergency phone lines for people to contact if they develop signs of the illness. The sessions in court were postponed for a few days [10]. Till date the scenario is as follows: Total cases 2.72 Cr + 2.09 L, Recovered 2.44 Cr + 2.96 L, Deaths 3.11 L + 4,157.

11.6   THE SECOND WAVE

Though the year 2021 bought a new ray of hope in the form of vaccine, there was no confirmation of when it was supposed to be applied. In March 2021, there was a certain outbreak of COVID-19, the second wave which came up with different symptoms and mild and severe ones. Though the testing game had improved from the first wave of COVID-19 the increased cases of COVID-19 could not be controlled in this phase and approximately 5,000 deaths (approx.) were reported on May 19th, 2021. The new wave came up with a mutated version of corona spreading quicker than before and cumulating the number of cases too.

New developed symptoms include instant saturation of oxygen level without any other symptoms of the earlier virus affecting ages between 18 and 60 years of age. And like before, the more prone to the severity of the cases is of the patients who have other auto-immune diseases, diabetes, asthma patient, cancer, etc. With the death toll rising above normal the Government couldn't handle the scenario, Maharashtra had seen the worst phase of the second wave cases. Figure 11.7 shows the affected in the 1st and 2nd wave.

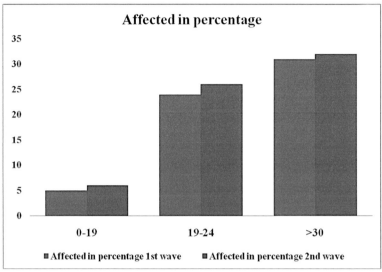

FIGURE 11.7   Affected in 1st and 2nd wave.

With an increase in the number of instances, people's need for vaccinations has grown as well, resulting in a vaccine supply shortage. Given India's population, widespread production of the vaccine is critical. Many people lost their employment as a result of the epidemic, as seen in Figure 11.8. Many jobholders lost their jobs due to this pandemic; Figure 11.8 shows the data.

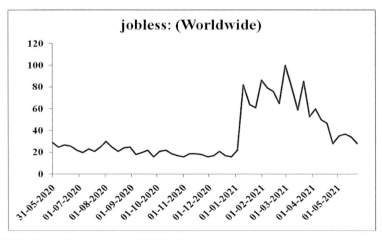

FIGURE 11.8   Job loss due to COVID-19.

11.7   IN OTHER COUNTRY

The first woman to be tested positive in South Korea was a Chinese woman of 30 years of age on 20th January 2022 and later, a 55-year-old man was confirmed as the second case of COVID-19 who worked at Wuhan and returned to South Korea. Till 17th February 2020 around 30 cases of COVID-19 cases were found in South Korea [9]. South Korea noted 79.4% of coronavirus cases were confirmed. Cases rising and dropping in April till September, October saw a drop in the number of cases with 100 cases per day [9].

The month of September also witnessed a slight drop in new cases so they uplifted the restrictions on few places like restaurants would open keeping new restrictions in mind, after school activities to restart, also gyms and bars to reopen keeping in mind the restrictions on social distancing, masking, sanitizing, etc., but as November arrived there was again a surge in the cases closing all gyms and bars. December reports the onset of the third wave of virus which was bought about by a family of three who returned from the United Kingdom to South Korea.

On 10 February 2021, South Korea declared its first approval of a COVID-19 vaccine to Oxford–AstraZeneca knocking a warning for the senior citizens due to limited intervention [9]. But that was not the ultimatum for South Korea as they saw a flood of the fourth wave in March.

South Korea, like any other country has suffered economically, but they have allowed the small firm business to run at its own pace. Schools and colleges are continuing with internet learning. And restrictions on movements have been removed but norms of social distancing, wearing masks in public remains constant.

The first two positive cases in Italy were evidenced in the body of two Chinese tourists in Rome who have traveled from China on 31st January 2020 [10]. By the end of March, there was a cluster of cases all over Italy. Italy reported 3,83,854 active cases as of 9th May 2020, one of the highest COVID cases in the world. According to the Italian National Institute of Statistics (ISTAT), Italy has a mortality rate of 2,036 deaths per million. According to ISTAT, the mortality rate will increase by 1,00,526 fatalities in 2020, compared to the previous five years [10].

One of the major effects of COVID-19 is going digital, online classes, online jobs, work from home. So, the new technologies are there in Figure 11.9.

On 4th March 2020, the region was closed down for a few weeks after the country reached 100 deaths in a day. On 24th March there was an extension of lockdown putting more constraints on movement and gatherings till 13th April. There was an announcement of schools and colleges being shut down till September.

In and around June there was leniency on the lockdown letting people freely move in the region. Though several charges will be applied if rules of social distancing, masks were not maintained.

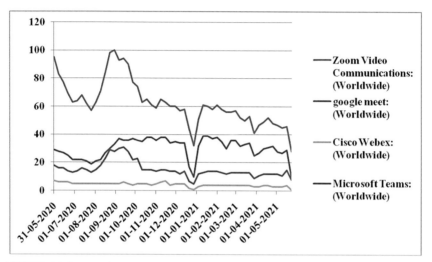

FIGURE 11.9 Popular online platforms.

11.8 EXPONENTIAL FORECASTING

The model equation is as follow:

$$z(t) = \chi(t) + \phi(t) \tag{11.1}$$

where; $z(t)$ is a random variable that describes the effect of stochastic fluctuation and takes a constant at time t and may change slowly over time; is a random variable that takes a constant at time t and may change slowly over time; (t) is a random variable that takes a constant at time t and may change slowly over time.

The core notion of an exponential smoothing model is that it will learn a bit from the most recent demand observation and remember a bit from the last prediction it did at each period. The model's most recent forecast includes both a component of the previous demand observation and a component of the prior projection. Based on historical demand, this past projection contains all the model has learnt thus far. The learning rate, alpha, determines the significance of the most recent demand observation. Let's take a mathematical look at it [12, 13].

$$f_{t+1} = az_t + (1-a)f_t$$
$$0 \prec a \leq 1 \tag{11.2}$$

Alpha, which is a ratio, measures the relevance of the most current observation in contrast to the relevance of demand history.

Alpha dt–1 is the previous demand observation multiplied by the learning rate. You may argue that the model lends considerable weight to the most recent demand incidence (alpha).

The amount of information recalled by the model from its previous forecast is (1-alpha) ft–1. Because ft–1 was defined as partially dt–2 and ft–2, the recursive magic occurs here. There is a natural trade-off to be made here between learning and remembering, between being reactive and being steady.

Once we've departed the historical era, we need to construct a prediction for future periods. It's simple: the most recent projection is simply projected ahead. If we define f_t^* as the most recent projection based on past demand.

$$f_{t \succ t^*} = f_t^* \tag{11.3}$$

- As F_1 is not known, let $f_1 = z_1$
- Take the average of the first few observations in the data series for the first smoothed value.
- The smoothing constant is a chosen value between zero and one (0 and 1).

Using the Eqn. (2):

$$f_{t+1} - f_t = a(z_t - f_t) \tag{11.4}$$

what is,

$$f_{t+1} = (f_t + ar_t) \tag{11.5}$$

where; residue $t_t = (z_t - f_t)$ is the error during forecast for t.

From iteration on Eqn. (2):

$$f_1 = z_1 \tag{11.6}$$

$$f_2 = a.z_1 + (1-a)f_1 = z_1 \tag{11.7}$$

The generalized form is [14]:

$$f_{t+1} = a\sum_{k=0}^{t-1}(1-a)^k.z_{t-k} + (1-a)^t z_1, t \in \Re \tag{11.8}$$

11.9 LIMITATION

Nonetheless, the surgery is strongly advised. However, there are several limitations:

- It offers no predictions regarding what will happen in the future. Our second model, exponential averaging with trend, also known as the double multiple regression, will handle this issue.
- It is unaffected by seasonal changes. The triple exponential smoothing strategy will be used to overcome this issue.
- It would be unable to use any data sources.

11.10 RESULT ANALYSIS

The results are as in Figures 11.10 and 11.11.

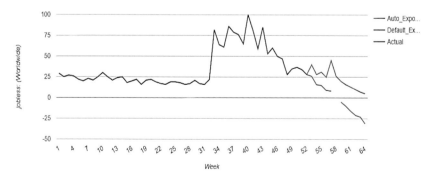

FIGURE 11.10 Jobless due to COVID-19 with forecasting.

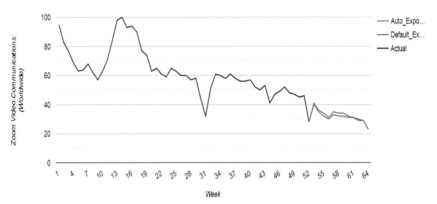

FIGURE 11.11   Use of Zoom in COVID-19 with forecasting.

Figures 11.10–11.14 show the real data used (worldwide) in blue color, green shows the forecast using default exponential furcating algorithm, and orange color show forecasting using the auto exponential algorithm of jobless, using zoom, google meet, Cisco Webex, and Microsoft Teams. Figure 11.15 shows the status of vaccination in India with forecasting.

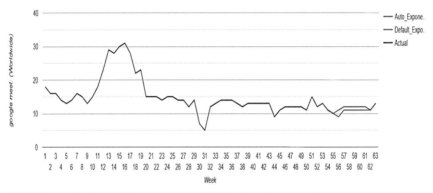

FIGURE 11.12   Use of Google meet in COVID-19 with forecasting.

11.11   VACCINATION

The immunization began on December 27, 2020, with the public receiving more than 9,000 doses of the Pfizer–BioNTech vaccine [10]. These dosages were initially given to hospital medical and health workers. Lazio was the

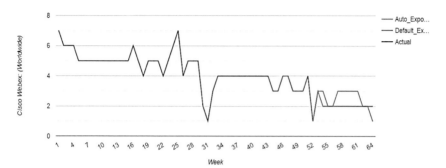

FIGURE 11.13   Use of Cisco Webex in COVID-19 with forecasting.

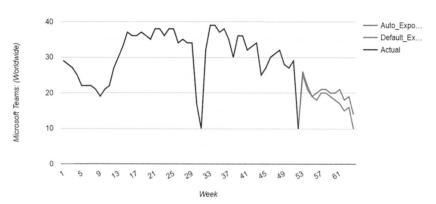

FIGURE 11.14   Use of MSTeams in COVID-19 with forecasting.

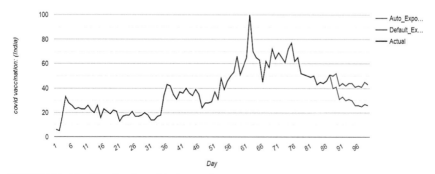

FIGURE 11.15   Vaccination with forecasting.

first area to begin immunizations, which took place in the Spallanzani Hospital in Rome.

The European EMA approved the Moderna vaccine on January 6, 2021, and millions of individuals have already got it as of May 3rd.

Nature's most dangerous gift is the COVID-19, in the 1st wave it has come with the Amphan cyclone, and in the 2nd wave, it comes with the Yash cyclone. Where it is said that the main point is to stay home, but these cyclones destroyed many houses and many of the people are out of home, for shelter they have to stay in a place where to maintain the social distance is itself a challenge. Figure 11.16 shows the vaccination status in India and worldwide for the last 90 days. As per the latest update, only 11.2% of people got their 1st vaccine and only 3.1% of people got the 2nd dose of vaccination in India.

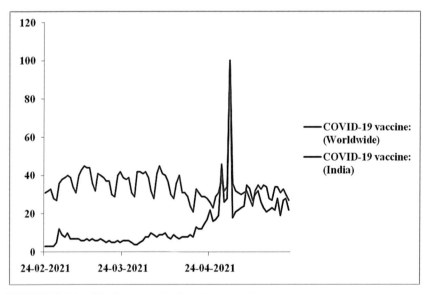

FIGURE 11.16   Vaccination in the last three months.

11.12   CONCLUSION

It is very clear from the data analysis and data forecasting that, the spread of COVID19 is decreasing day by day and the chance of getting the vaccine is also increased but slowly. The population is one of the major challenges. Apart from this, the use of the online teaching-learning process, technology, IT is rapidly growing nowadays.

ACKNOWLEDGMENTS

The authors are highly obliged to the Department of Education, Institute of Education for Women, Hastings, Kolkata, India, and Arya Parishad Vidyalaya for Girls, Kolkata, India for their constant support and moral help. Though the work is not supported by any Foundation, the laboratory of the institute was helped to do the work smoothly. We thank our friends who provided insight and expertise that greatly assisted the research, although they may not agree with all of the interpretations of this chapter.

KEYWORDS

- **COVID-19**
- **data analysis**
- **exponential algorithm**
- **forecast**
- **information technologies**
- **middle east respiratory syndrome**
- **severe acute respiratory syndrome**

REFERENCES

1. Mayo Foundation for Medical Education and Research. https://www.mayoclinic.org/ (accessed on 11 October 2022).
2. Zoumpourlis, V., Goulielmaki, M., Rizos, E., Baliou, S., & Spandidos, D. A., (2020). [Comment] The COVID-19 pandemic as a scientific and social challenge in the 21st century. *Molecular Medicine Reports, 22*(4), 3035–3048.
3. World Health Organization, (2020). *Coronavirus Disease 2019 (COVID-19): Situation Report*, 72.
4. Lima, C. M. A. D. O., (2020). Information about the new coronavirus disease (COVID-19). *Radiologia Brasileira, 53*(2), V, VI.

5. Lau, H., Khosrawipour, V., Kocbach, P., Mikolajczyk, A., Ichii, H., Schubert, J., & Khosrawipour, T., (2020). Internationally lost COVID-19 cases. *Journal of Microbiology, Immunology and Infection, 53*(3), 454–458. World Health Organization, https://www.who.int (accessed on 11 October 2022).

6. Andrews, M. A., Areekal, B., Rajesh, K. R., Krishnan, J., Suryakala, R., Krishnan, B., & Santhosh, P. V., (2020). First confirmed case of COVID-19 infection in India: A case report. *The Indian Journal of Medical Research, 151*(5), 490.

7. Business Today, *First Case of Coronavirus Confirmed in India; Student Tested Positive in Kerala.* https://www.businesstoday.in (accessed on 11 October 2022).

8. Gupta, R., Pal, S. K., & Pandey, G., (2020). *A Comprehensive Analysis of COVID-19 Outbreak Situation in India.* MedRxiv.

9. Shin, H., & Cha, S., (2020). *Like a Zombie Apocalypse: Residents on Edge as Coronavirus Cases Surge in South Korea* (p. 20). Thomson Reuters. Archived from the Original, 20.

10. Severgnini, C., (2020). Coronavirus, first two cases in Italy 'They are two Chinese on vacation in Rome' They arrived in Milan. *Corriere Della Sera, 31.*

11. Medical News Today. *COVID-19: Latest News and Resources.* https://www.medicalnewstoday.com (accessed on 11 October 2022).

12. Hyndman, R., Koehler, A. B., Ord, J. K., & Snyder, R. D., (2008). *Forecasting with Exponential Smoothing: The State Space Approach.* Springer Science & Business Media.

13. Gardner Jr, E. S., (1985). Exponential smoothing: The state of the art. *Journal of Forecasting, 4*(1), 1–28.

14. Ostertagova, E., & Ostertag, O., (2012). Forecasting using simple exponential smoothing method. *Acta Electrotechnica et Informatica, 12*(3), 62.

15. Towards Data Science, https://towardsdatascience.com (accessed on 11 October 2022).

CHAPTER 12

Comparative Analysis of SVM and KNN in Breast Cancer Classification

SOUMYAJIT PAL

Faculty of Information Technology, St. Xavier's University, Kolkata, West Bengal, India

ABSTRACT

Cases of cancer in various forms, especially breast cancer among women, have been on the rise in recent times. If we are able to correctly categorize patients having malignant or benign form of breast cancer on the basis of historical data, it would not only eliminate human errors that often get introduced in proper identification of diseases but would also go a long way in providing correct diagnosis and ensure speedy recovery. Although there are a number of classification algorithms available to carry out this task on a given dataset, this chapter focuses on support vector machine (SVM) classifier and K-nearest neighbor classifier. Both techniques are applied on given data and results are recorded. This is followed by a comparative analysis of both the algorithms in achieving the task at hand using appropriate parameters. The chapter concludes with an overview of suggestions on how to obtain better classification results and the scope of reliance on data to arrive at concrete results for the diagnosis of patients suffering from critical ailments. The Breast Cancer Wisconsin (Diagnostic) Data Set is used for experimentation purposes.

System Design Using the Internet of Things with Deep Learning Applications.
Arpan Deyasi, Angsuman Sarkar, and Soumen Santra (Eds)
© 2024 Apple Academic Press, Inc. Co-published with CRC Press (Taylor & Francis)

12.1   INTRODUCTION

The menace of cancer has engulfed the lives of millions around the world. It is a disease which oftentimes does not exhibit any initial symptoms but reveals its deadliness at a time when there is little left to be done. Had there been a reliable way to detect cancer at an early stage, saving the lives of people would have been a lot easier. This manuscript takes a step forward in that very direction by trying to test and compare the performance of two popularly used machine learning models – K-nearest neighbors (KNN) and support vector machine (SVM) in trying to classify tumors as benign or malignant using the Wisconsin Breast Cancer Data Set [1]. Breast cancer is particularly common among women but it has been found to be curable [2] in approximately 70% to 80% of patients provided there is early detection. According to Ref. [2], India is particularly seeing a rising number of cases of breast cancer among younger women below the age of 40. The trend becomes even more disturbing when it is compared to data 25 years ago. The cancer detection problem is one which requires immediate attention. According to Ref. [3], the number of new cases in males and females will rise to 0.934 million and 0.935 million respectively by 2026. The study [3] also goes on to state that 50% to 60% of total patients will be diagnosed with either tobacco-related cancer, breast cancer or cervical cancer. Keeping the grim statistics aside, accurate prognosis based on historical data can go a long way in understanding the symptoms that may lead to cancer at a later stage in a person's life and thereby preventing the onset of the disease.

12.2   EXPERIMENTAL METHODS

Given historical data of patients, the task is to classify an unknown data point expressed as a set of values of various features of a digital image of cell nuclei as either malignant tumor or benign tumor. Since class labels are available for training the models, this is a case of supervised learning. As class labels are categorical (Malignant and Benign), it is a binary classification problem.

12.2.1   K-NEAREST NEIGHBORS (KNN)

The K-nearest neighbors (KNN) algorithm [4] is an instance-based learning algorithm where data is merely stored during the training phase but not used to build a model. In the testing phase, all the data is used for classifying an unknown point. The 'K' refers to the number of neighbors which participate in the decision-making process of classification. There are mainly three steps to the algorithm:

➢ **Step 1:** Calculating the distance from the unknown point, P, to all points in the dataset.
➢ **Step 2:** Choosing 'K' points which have minimum distance.
➢ **Step 3:** Classifying P based on majority vote.

The main challenge lies in choosing the value of 'K.' There is no fixed method based on which 'K' can be chosen. It largely depends on the data set being worked on. Oftentimes, various values of 'K' are tried out. The value which gives the best performance is chosen. A rule of thumb is to choose an odd number when the count of class labels is even. Euclidian's distance formula is most often used for finding how far the unknown point is from all points in the dataset. Other measures such as Minkowski distance or Cosine similarity may be used. In the proposed work, Euclidean distance is used because the number of dimensions is not very large.

12.2.2   SUPPORT VECTOR MACHINE (SVM)

Support vector machine (SVM) is a supervised learning algorithm [5] used for solving both classifications as well as regression problems. For classification tasks, it works by fitting the best hyperplane to segregate the data points into the given class labels. In case the points are not linearly separable, it employs the kernel trick method to elevate the points in higher dimensions, thereby being able to make a distinction among the various data points. There is particular emphasis on finding out the support vectors – points closest to the hyperplane. Support vectors play a crucial role by calculating the margin and defining the separating line better.

12.2.3   THE DATASET

12.2.3.1   DATA COLLECTION

The Wisconsin Breast Cancer Data Set [1] has records of 569 patients. Each row of data is represented using features (columns). The process of data collection [6] could be summarized using the following steps:

➢ **Step 1:** Patient with tumor was recognized.

- ➤ **Step 2:** Using fine needle aspiration (FNA), cells of breast mass were obtained (aspirant).
- ➤ **Step 3:** Digital image f(x, y) of resolution 640 x 400 of aspirant was found.
- ➤ **Step 4:** f(x, y) was fed to an image analysis program [7–9] which used curve fitting to obtain the boundary portion of the nuclei of the cells marked manually with dots using mouse pointer.
- ➤ **Step 5:** 10 features were computed for each nucleus found in the previous step – area, radius, perimeter, symmetry, number of concavities, size of concavities, fractal dimension of the boundary, compactness, smoothness, and texture.
- ➤ **Step 6.1:** Mean of each of the 10 features was calculated.
- ➤ **Step 6.2:** Standard Error of each of the 10 features was calculated.
- ➤ **Step 6.3:** Extreme Value of each of the 10 features was calculated.
- ➤ **Step 7:** Every patient's unique identification number was added to the dataset.
- ➤ **Step 8:** The target attribute denoting malignant or benign tumor was also added.
- ➤ **Step 9:** Repeatedly steps 1 to 8 were carried out for every patient.

12.2.3.2   *DATA CLEANING*

Gaining an insight into the process of data collection adds to domain knowledge which is essential for data analysis, thus leading to the gathering of useful information and patterns. On close inspection, an attribute by the name of Unnamed: 32 are found. It contains only null values. Such an attribute has no role to play in the prediction process and is removed. The unique identification number of a patient is dropped because it adds no value to the classification of tumors. The target attribute diagnosis contains exactly two unique values – M (for malignant) and B (for benign). Out of the 569 patients' tumor examination, 62.74% was found to be benign, while the remaining 37.26% was found to be malignant.

From the count plot of Figure 12.1, it is clearly understood that the given dataset is not perfectly balanced. It can be considered partially balanced. The target variable diagnosis contains alphabetic labels such as M and B, which are not understood by machine learning algorithms. Thus, they are mapped into numeric values: 1 for M and 0 for B.

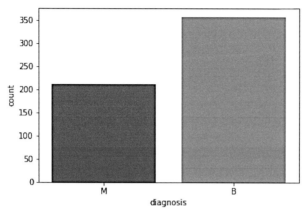

FIGURE 12.1   Count plot of diagnosis attribute containing 357 benign cases and 212 malignant cases.

Missing values in attributes pose problems and must be inspected for in the dataset. If present in large numbers, they have to be dealt with using standard procedures. However, the Wisconsin Breast Cancer Dataset contains zero missing values. Finding the correlation among the attributes can prove to be significant as features which are highly correlated represent redundant information and may be removed from the dataset completely. This task is particularly important because of the presence of a large number of variables, where each variable represents a dimension. Having to deal with a higher number of variables leads to the problem of Curse of Dimensionality, where running times approach exponential and error rate increases. Separate scatter plots on attributes containing mean values, attributes containing standard error values and attributes containing worst values give clarity on how those features are correlated.

Analyzing the scatter plots reveals almost perfect linear correlation among attributes mean area, mean perimeter and mean radius. There appears to exist linear correlation among attributes concavity mean, mean of number of concave points and mean compactness as well. This presents the problem of multi-co-linearity among these independent variables. It is resolved by choosing only one of these features for model building.

The observations made on the scatter plot are verified using a correlation matrix. A threshold value of 0.90 is fixed, and if the correlation between two features lies above it, one of the features is dropped. For instance, the mean radius column has a correlation of 1 and 0.99, respectively with mean perimeter and mean area, respectively. This proves that there exists multi-co-linearity among these features, i.e., they contain redundant information. It would be prudent to choose only one of these features going into further analysis. The mean radius column is chosen and the remaining two are dropped. Also, features worst radius, worst perimeter and worst area have a correlation value above the chosen threshold when compared with mean radius and are hence removed from the dataset. The worst texture attribute is dropped and the mean texture is retained. Similarly, attributes such as mean concavity, worst number of concave points, standard error on area and standard error on perimeter are removed from the dataset. Ultimately, it comes down to 21 columns.

The output class label diagnosis is separated from the independent attributes such as mean radius, worst value of fractal dimension, standard error on smoothness among others (Figure 12.2).

```
Data columns (total 21 columns):
 #   Column                  Non-Null Count   Dtype
---   ------                  --------------   -----
 0   diagnosis               569 non-null     int64
 1   radius_mean             569 non-null     float64
 2   texture_mean            569 non-null     float64
 3   smoothness_mean         569 non-null     float64
 4   compactness_mean        569 non-null     float64
 5   concave points_mean     569 non-null     float64
 6   symmetry_mean           569 non-null     float64
 7   fractal_dimension_mean  569 non-null     float64
 8   radius_se               569 non-null     float64
 9   texture_se              569 non-null     float64
10   smoothness_se           569 non-null     float64
11   compactness_se          569 non-null     float64
12   concavity_se            569 non-null     float64
13   concave points_se       569 non-null     float64
14   symmetry_se             569 non-null     float64
15   fractal_dimension_se    569 non-null     float64
16   smoothness_worst        569 non-null     float64
17   compactness_worst       569 non-null     float64
18   concavity_worst         569 non-null     float64
19   symmetry_worst          569 non-null     float64
20   fractal dimension worst 569 non-null     float64
```

FIGURE 12.2   Final list of columns on which prediction is performed.

12.2.3.3   *DATA PREPARATION FOR MODEL BUILDING*

The dataset is split into train and test sets such that 30% of data is kept aside for testing while the remaining 70% of data is used for training.

Since the attribute values are in different scales, standardization is applied to all the features so that no feature has a greater effect on the model than others.

12.3   RESULTS AND DISCUSSION

Four models are prepared separately – Model M_1 (trained using KNN where K = 7); Model M_2 (trained using KNN, where K = 5); Model M_3 (trained using KNN, where K = 3); and Model M_4 (trained using SVM). The four models are tested separately and the results are summarized in Table 12.1. The metrics used for evaluating the models are precision, recall, F_1-score (F-score) and accuracy.

$$Precision = \frac{True\ Positive}{True\ Positive + False\ Positive} \tag{12.1}$$

$$Recall = \frac{True\ Positive}{True\ Positive + False\ Positive} \tag{12.2}$$

$$Accuracy = \frac{True\ Positive + True\ Negative}{True\ Positive + True\ Negative + False\ Positive + False\ Negative} \tag{12.3}$$

$$F_1 = \frac{True\ Positive}{True\ Positive + \frac{1}{2}(False\ Positive + False\ Negative)} \tag{12.4}$$

Precision measures the rate of correct positive identifications. Recall measures the rate of actual positives identified correctly. Accuracy gives an idea as to how close the predicted result is to the actual result. F measure takes into account both false positives and false negatives, thereby giving an overall weighted mean of precision and recall. These values are computed from the confusion matrix shown in Figure 12.3.

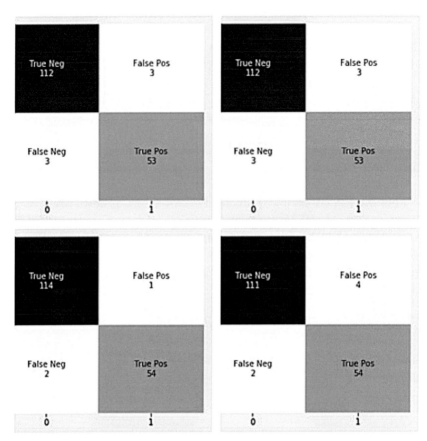

FIGURE 12.3   Confusion matrix of models M_1 (top left); M_2 (top right); M_3 (bottom left); and M_4 (bottom right).

TABLE 12.1   Comparative Performance of the Models

Model Name	Number of Nearest Neighbors (K)	Average Precision	Average Recall	Average F-Score	Accuracy
M_1	7	0.96	0.96	0.96	0.965
M_2	5	0.96	0.96	0.96	0.965
M_3	3	0.98	0.98	0.98	0.982
M_4	–	0.97	0.96	0.97	0.965

On evaluating the models, it is found that M_3 (KNN with K = 3) performs the best on the given dataset and has the highest F measure. The Support Vector Classifier seems to be doing better than KNN (for K = 5 or K = 7).

12.4 CONCLUSION

Since this manuscript solely focuses on two algorithms – SVM and KNN – there was no scope to try out other classification models such as Logistic Regression, Random Forest, Decision Tree, among others. In search of better performance, the above methods could also be experimented with. In the end, it is about minimizing the human errors that get introduced during medical procedures. Proper prognosis based on quality data can go a long way in the fight against cancer in general and breast cancer in particular.

KEYWORDS

- **fine needle aspiration**
- **K-nearest neighbors**
- **learning algorithm**
- **machine learning**
- **random forest**
- **support vector machine**

REFERENCES

1. Murphy, P. M., & Aha, D. W., (1992). *UCI Repository of Machine Learning Databases.* https://archive.ics.uci.edu/ml/datasets/breast+cancer+wisconsin+%28original%29 (accessed on 11 October 2022).
2. The Pink Initiative, (2020). *Trends of Breast Cancer in India.* breastcancerindia. net. [Online]. Available: https://www.breastcancerindia.net/statistics/trends.html (accessed on 11 October 2022).

3. D'Souza, N. D., Murthy, N. S., & Aras, R. Y., (2013). Projection of cancer incident cases for India -till 2026. *Asian Pac J Cancer Prev., 14*(7), 4379–4386. PMID: 23992007.
4. Cunningham, P., & Delany, S., (2007). k-Nearest neighbor classifiers. *Mult. Classif. Syst.*
5. Evgeniou, T., & Pontil, M., (2001). *Support Vector Machines: Theory and Applications, 2049*, 249–257.
6. Mangasarian, O. L., Street, W. N., & Wolberg, W. H., (1995). Breast cancer diagnosis and prognosis via linear programming. *Operations Research, 43*(4), 570–577.
7. Wolberg, W. H., Street, W. N., & Mangasarian, O. L., (1993). Breast cytology diagnosis via digital image analysis. *Analytical and Quantitative Cytology and Histology.*
8. Wolberg, W. H., Street, W. N., & Mangasarian, O. L., (1995). Image analysis and machine learning applied to breast cancer diagnosis and prognosis. *Analytical and Quantitative Cytology and Histology, 17*(2), 77–87.
9. Street, W. N., Wolberg, W. H., & Mangasarian, O. L., (1993). Nuclear feature extraction for breast tumor diagnosis. *IS&T/SPIE 1993 International Symposium on Electronic Imaging: Science and Technology, 1905*, 861–870. San Jose, CA.

Index